LEARN 51 TRICKS OF MULTIPLICATION IN ONE DAY

FOR ALL AGE LEARNERS
&COMPETITIVE
EXAMINATIONS

DR. BINAYAK SAHU

PUBLISHED BY
UAB PUBLICATION HOUSE, ODISHA

Copyright © 2020 Dr.Binayak Sahu.

All rights reserved.

No part of this publication may be reproduced, stored in a retrieval system or transmitted, in any form or by any means, electronic, mechanical, photocopying, recording or otherwise without prior permission of the Publisher.

DEDICATION

In Bhagbad-GITA
Lord Krishna tells:

Your right is to work only
But never to its fruits;
Let not the fruits of action be thy motive,
Not let thy attachment be to inaction

Dedicated
In the Loving Memory
Of my beloved father
Late Suryanarayan Sahu

CONTENTS

    PREFACE

1   CHAPTER ONE   Page 1

2   CHAPTER TWO   Page 12

3   CHAPTER THREE   Page 29

4   CHAPTER FOUR   Page 41

5   CHAPTER FIVE   Page 59

6   CHAPTER SIX   Page 72

PREFACE

The book titled as Learn 51Tricks of Multiplication in One Day is meant for students of all ages across schools, colleges& Universities as well as for those, who prepare for various competitive examinations.

The tricks of multiplication for different set of numbers are explained to help the students to solve the problems of multiplications within a shortest time possible. All the four arithmetic operations, namely, addition, subtraction, multiplication and division, individually has one's own characteristics & importance. This book is an effort to highlight the tricks and techniques to educate or inform all its readers the possibilities of the shortest ways to conduct any multiplication. Mathematics is a game of numbers. Numbers contain digits 1 – 9 and 0. If you love these numbers, then numbers start loving you. The learners will find it interesting to explore the possibilities in performing any multiplication. Proper examples are solved to aid it's learners to help self in learning the tricks suitably. It is sincerely hoped that this book will satisfy the common needs of the learners on the subject of multiplication.

I do not claim any originality but the presentation of the subject content is mine & unique. I do express here with my indebtedness to my teachers, authors and students who inspired me to prepare this book.

I am very much thankful to the UAB Publishing House for their great efforts in editing & publishing this book. I am also thankful to my mother, wife and son for their patience & support during preparation of this text.

Suggestions for improvement of this book will always be accepted with thanks.

Dr. Binayak Sahu

CHAPTER ONE

1. BASICS

Multiplication: It is a rule or operation by which quantities are multiplied.

Multiplicand: It is a quantity to be multiplied by another quantity.

Example:3 X 5 *is read as* 3 *multiplied by* 5. Here "3" is the multiplicand

Example:3 X 5 *is read as* 5 *multiplied by* 3. Here "5" is the multiplicand

NOTE: Multiplication is also termed as "times addition of". Multiplication is often described as repeated addition.

Example:3 X 5 *can be meant as* 3 *times addition of* 5. That is **5 + 5 + 5 = 15** (5 is added 3 times)

Example:5 X 3 *can be meant as* 5 *times addition of* 3. That is **3+3+3+3+3 = 15** (3 is added 5 times)

Multiplication – a mathematical operation where a number is added to itself a number of times. Multiplicand – is the number being multiplied and Multiplier – is the number doing the multiplying. The answer is called the product or the multiple.

NOTE: The multiplicand and the multiplier are both factors of the product or the multiple.

1.1 MULTIPLICATION SIGN & ITS USES:

The Multiplication sign, also known as 'times sign' or the 'dimension sign', is the symbol 'x'. It looks exactly the same under a certain degree of rotation. 'x' is said to satisfy the property of radial or rotational symmetry. Wikipedia lists the various uses of this multiplication sign and other symbols of multiplication used across subjects.

In mathematics, the symbol 'x' has a number of uses, including:

- Multiplication of two scalar numbers, where it is read as 'times 'or 'multiplied by'.

- Cross product of two vectors, where it is usually read as 'cross'.
  Example, $\vec{a} \times \vec{b}$, where $\vec{a} \& \vec{b}$ are vectors.

- Cartesian product of two sets, where it is usually read as 'cross'. For example: A x B.

- Geometric dimension of an object, such as while expressing the size of a room, it is 10 feet x 12 feet in area, where it is usually read as 'by'(10 feet by 12 feet).

- Dimensions of a matrix, where it is usually read a 'by'. A_{mxn} is a m x n matrix, with 'm' rows and 'n' columns.

- A statistical interaction between two explanatory variables, where it is usually read 'by'.

- In biology, the multiplication sign is used in a botanical hybrid name.
- The multiplication sign is also used by historians for an event between two dates. When employed between two dates, for example 1225 and 1232, 1225x1232 means 'no earlier than 1225 and no later than 1232'.
- In algebra, a multiplication symbol is usually omitted and 'a multiplied by b' can be simply written as ab.
- In many countries, rather than 'x', the primary symbol for multiplication is '·', the dot operator as in 'a·b'.
- Computer language syntax was restricted to the ASCII character set, and the '*' asterisk became the de facto standard notation of the multiplication operator in computing.
- A mathematical operation, symbolized by axb, a · b, a*b, ab, are the representations of multiplication between the quantities a and b.

"Multiplication is the process of calculating the total of one number multiplied by another"

1.2 BASE:

The numbers 10, 100, 1000, 10000 etc. are treated as bases for its nearby numbers. Whereas working bases are declared sometimes to ease a multiplication.

Example: The numbers 7, 8, 9, 11, 12, 13 . . . .are considered as near to the base 10.

Whereas, 91, 92. . . 97, 99, 101, 102. . .109, etc. are near to the base 100.

Similarly, numbers like 1001, 1005, 1009, 999, 997, 991,. . . are near to the base 1000.

NOTE: We can assume such bases as required. Do remember, multiple bases can be assumed for a problem.

1.3 COMPLEMENT

If sum of two digits is equal to 10, then each digit is said to be a complement of the other from 10.

Example: 7 + 3 = 10. Here 7 is the complement of 3 from 10 and vice versa.

NOTE: This concept of complement can also be extended from any base.

Example: 77 + 23 = 100. Here 77 is the complement of 23 from 100 and vice versa.

Example: 981 + 19 = 1000. Here 981 is the complement of 19 from 1000 and vice versa.

NOTE: The reader is hereby advised to focus on the complements from 10 and 9 only, unless declared otherwise.

1.4 COMPLEMENT TABLE

Complements from 10

Digit	1	2	3	4	5	6	7	8	9
Complement	9	8	7	6	5	4	3	2	1

NOTE: Digits in each column added together makes 10.

Complements from 9

Digit	0	1	2	3	4	5	6	7	8
Complement	9	8	7	6	5	4	3	2	1

NOTE: Digits in each column added together makes 9.

Complements from 100

Digit	11	20	- -	66	43	99
Complement	89	80	- -	34	57	01

NOTE: Digits in each column added together makes 100.

Complements from 1000

Digit	101	127	-	23	943	7
Complement	899	873	-	977	57	993

NOTE: Digits in each column added together makes 1000.

1.5 NOTATION OF COMPLEMENT

$1\bar{2}$, here the bar on 2 indicates complement.

Explanation: $1\bar{2} = 10 - 2 = 8$

Example: $3\bar{4} = 30 - 4 = 26$

Explanation: The number 4 is with bar and is in the units place. 3 is in the ten's place. So, 4 is subtracted from 30.

Example: $2\bar{4} = 20 - 4 = 16$

Example: $13\bar{6} = 130 - 6 = 124$

Fifty Tricks of Multiplication

Explanation: The number 6 is with bar and is on the units place. Here we treat either 13 is in the ten's place / 1 is in the hundred's place & 3 in the ten's place and in either case '13' represent 130. So, 6 is subtracted from 130.

Example: $\bar{4} = 0 - 4 = -4$

Explanation: The number 4 is with bar and is subtracted from 0.

Example: 27 *can be written as* $3\bar{3}(= 30 - 3 = 27)$

Example: $8\bar{4}1 = 800 - 41 = 759$

Explanation: The number 41 is with bar and 8 is in the hundred's place. So, 41 is subtracted from 800.

Example: $2\bar{2}3 = 200 - 23 = 177$

Example: $7\bar{4}6 = 706 - 40 = 766$

Explanation: The number 4 is with bar and is on the ten's place. The number 7 is in the hundred's place & 6 in the unit's place. So, 40 is subtracted from 706.

Example: $7\bar{4}6 = 700 - 46 = 754$
Example: $1\bar{6}9 = 109 - 60 = 49$
Example: $3\bar{8}65 = 3005 - 860 = 2145$
Example: $\bar{6}9 = 0 - 69 = -69$
Example: $27\bar{2}1 = 2700 - 21 = 2679$

Explanation: The number 21 is with bar and 27 is in the hundred's place. So, 21 is subtracted from 2700.

1.6 ALL FROM 9 & LAST FROM 10

To find the complement of a number from a base we need not subtract the number from the base every time. We can use the Vedic Sutra "All from 9 and the last from 10", known as Nikhilam Sutra, while calculating the complement of a given number.

Example: Find the complement of number 167 and base 1000.

Applying all from 9, 1 from 9 is 9 − 1 = 8
6 from 9 is 9 − 6 = 3
And last from 10,   7 from 10 is 10 − 7 = 3
Therefore, 833 is the complement of 167 from 1000.

NOTE: 167 + 833 = 1000

Example: Find the complement of number 23 and base 1000. Note: 1000 has 3 zero's.

We write the number 23 as 023

Applying all from 9,   0 from 9 is 9 − 0 = **9**

2 from 9 is 9 − 2 = 7

And last from 10,   3 from 10 is 10 − 3 = 7

Therefore, 977 is the complement of 23 from 1000 and vice versa.

NOTE: 23 + 977 = 1000

Example: Find the complement of number 54 and base 100.

Note: 100 has 2 zero's.

We write the number 54 as 54 only.

Applying all from 9,   5 from 9 is 9 − 5 = 4

And last from 10,   4 from 10 is 10 − 4 = 6

Therefore, 46 is the complement of 54 from 100.

NOTE: 54 + 46 = 100

SUMMARY:
- (i) Two digits, if added together makes 10, 100, 1000 etc. then they are complements of each other.
- (ii) The number with a bar above it, indicates the complement of that number.
- (iii) We can find the complement of a number by the Vedic rule: "All from 9 and last from 10"(Here the word 'last' refers to the unit place.

PROPOSAL: It seems, readers can now compute the complement of any number from any base, mentally, with little more practice. Hope you do the best. Go ahead with examples of your own.

DO YOURSELF:

1. Find the complements of 23, 97, 7, 34, 39, 89, 77 from 100, 1000.

2. Find the complements of 123, 343, 444, 601, 100, 33, 71, 209, 999, 9 from 1000.

3. Write the values of $9\bar{6}9, 1\bar{4}, 22\bar{2}2, 1\bar{7}8$.

1.7 DOUBLING AND HALVING

Doubling a number means a number is to be multiplied by 2. Doubling a number 2 times, means a number is to be multiplied by 2, two times, and so on.

Example: Double 27 one time. We get 27 x 2 = 54

Example: Double 129 one time. We get 129 x 2 =

258.

Example: Double 22 two times. We get 22 x 2 = 44 and then 44 x 2 = 88.

Example: Double 139 three times. We get 139 x 2 = 278, 278 x 2 = 556 and finally, 556 x 2 = 1112.

Halving a number means a number is to be halved. Half a number once means, we make it into equal two parts. Similarly, we can half a number two times, three times etc.

Example: Half 28 one time. We get 14.

Example: Half 132 two times. We get 64, halving once, and then 32 after halving 2nd time.

Example: Half 1000 three times. We get 500, halving once, then 250 after halving 2nd time & finally it is 125.

Example: Half 13 two times. We get 6.5, halving once, and then 3.25 after halving 2nd time.

Hope, the operations of doubling and halving is well understood by the readers.

DO YOURSELF

1. Double the following numbers, one time, 2 times and 3 times:23. 133, 756. 155, 12.5, 23.4, 122.1, 75.5, 231 & 345.

2. Half the following numbers, one time, 2 times and 3 times:132,42, 266, 55, 228, 234, 222, 556, 232&347.

1.8 TEEN NUMBERS

The names of the numbers between 10 and 20 do

not follow the common rule and they are the sight words or numbers of mathematics. To ease the students in the early age to remember these numbers, it is said that the teen numbers are between 10 and 20 and they are eleven (11) through nineteen (19).

1.9 DIVISIBILITY

Divisibility by 2: If the unit place of a given number is divisible by 2.

Divisibility by 4: If the 1st two place of a given number is divisible by 4.

Divisibility by 3: If the sum of the digits of a given number is divisible by 3.

Divisibility by 5: If the unit place of a given number is either '5 or '0', then the number is divisible by 5.

Divisibility by 6: If the number is divisible by 2 & 3 then it is divisible by 6.

Divisible by 7: Using 7 create a zero at the unit place and kill it, repeat the process till you reach at a number which is either divisible by 7 else it is not divisible by 7.

Example: 343. 343 + 7 = 350. Kill '0'. We get 35, which is divisible by 7. Therefore 350 is divisible by 7.

Example: 3668. 3668– 28 (7 x 4=28) = 3640. Kill '0'. We get 364. 364 – 14(7 x 2 = 14) = 350. Kill the

'0'. We get, 35, which is divisible by 7. Therefore 3668 is divisible by 7.

Divisible by 8: If the 1st three places of a given number is divisible by 8.

Divisible by 9: If the sum of the digits of a given number is divisible by 9.

CHAPTER TWO

#Tr.1 MULTIPLICATION BY 2, 4, 8, 16:

The method we shall use for multiplications by 2, 4, 8, 16 ...is called the Double Method. We will use the process of doubling a number to find such multiplications by 2, 4, 6, 8 and so on.

Multiplication of any number by 2 is same as adding the number two times or otherwise doubling the number once.

- Multiplication by 2 means doubling the number.
- Multiplication by 4 means doubling the number 2 times.
- Multiplication by 8 means doubling the number 3 times.
- Multiplication by 16 means doubling the number 4 times and so on.

Example: 5 x 2 = 10, 5 is doubled one time.

Example: 32 x 2 (= 64), 32 doubled one time.

Example: 32 x 4 (= 128), 32 doubled 2 times. That is, 32 doubled to 64 & 64 doubled to 128.

Example: 125 x 8 (=1000), 125 doubled one time gives 250, 2nd time gives 500 and the third time it is 1000.

Example: 231 x 8 (=1000), 231 doubled one time gives 462, 2nd time gives 924 and doubled again, the third time it is 1848.

Fifty Tricks of Multiplication

Example: 713 x 8 (=1000), 713 doubled one time gives 1426, 2nd time doubling, it gives 2852 and doubled again, the third time it is 5704.

#Tr.2 **MULTIPLICATION BY 5, 25, 125:**

This multiplication involves the respective bases, namely, for multiplication by 5 we use 10, for 25 we use 100 and for 125 we use 1000.

Motivation 1:

$We\ can\ write,\ 5 = \dfrac{10}{2} = \dfrac{1}{2}(10)$

Method for: Any number to be multiplied by 5
Step 1: Multiply the number by 10
Step 2: Half the result once.
Alternatively, we can half the given number and then multiply the result by 10, to get the product.

Example: Find 123 x 5=615
Explanation: We multiply 123 by 10 & get 1230. Half 1230 once and get 615.

Example: Find 27 x 5=135 (27 x 10 = 270 & 270 halved is 135)

Example: Find 723 x 5=3615 (7230 is halved)

Example: Find 777 x 5= 3885 (7770 is halved)

Have some quick practice of multiplications of numbers by 5. Yes, it is multiply the given number by 10and then half it one time.

DO YOURSELF

1. Multiply 31, 342, 456, 671, 788, 989, 988, 717 by 5.

Motivation 2:

$$We\ can\ write,\ 25 = \frac{100}{4} = \frac{11}{22}(100)$$

Method for: Any number to be multiplied by 25.
Step 1: Multiply the number by 100
Step 2: Half the result two times.
Alternatively, we can half the given number two times & then multiply the result by 100, to get the product.

Example: Find 72 x 25 = 1800.
Explanation: We multiply 72 by 100 & get 7200. Half 7200 two times and get 1800.

Example: Find 271 x 25 = 6775. (27100 halved two times, gives 6775 (halved once: 13550, & then 6775).

Example: Find 102 x 25 = 2550. (10200 halved two times, gives 2550 (halved once: 5100, & then 2550).

Alternatively, half of 102 is 51, whose half is 25.5, multiplied by 100, gives 2550.

Example: Find 998 x 25 = 24950. (99800 halved two times, gives 24950 (halved once: 49900, & then 24950).

Alternatively, half of 998 is 449, and then halved is 249.5, multiplied by 100 gives 24950

Relax a bit, after multiplication by 5 now is the time to do some multiplications by 25. Yes all without using pen and paper. Do, directly in your mind.

DO YOURSELF

1. Multiply 13, 146, 460, 788, 485, 666 & 571 by 25.

Fifty Tricks of Multiplication

Motivation 3:

We can write, $125 = \dfrac{1000}{8} = \dfrac{111}{222}(1000)$

Method: Any number to be multiplied by 25
Step 1: Multiply the number by 1000
Step 2: Half the result three times.
Example: Find 66 x 125 = 8250.
Explanation: We multiply 66 by 1000 & get 66000. Half 66000 three times & get (33000, 16500) 8250.
Alternatively, having 66, three times, we get, 33, 16.5 and then 8.25, multiplied by 1000, we get 8250.
Example: Find 239 x 125 = 29875. (239000 halved three times, gives 6775 (halved once: 119500, halved 2nd time, gives 59750 & then 29875).
Example: Find 204 x 125 = 25500. (204000 halved three times, gives 25500 (halved once: 102000, halved 2nd time, gives 51000 & then 25500).

Relaxed enough, writing is easy, but it can be done. Yes go ahead with solving the multiplications by 125 in your mind.

DO YOURSELF

1. Multiply 11, 46, 762, 804, 413, 888, 234, 117, 3245, 4066, 609 &7740 by 125.

#Tr.3 MULTIPLICATION by 9 PATTERN:

9 Table

9 x 1 = 09
9 x 2 = 18
9 x 3 = 27
9 x 4 = 36
9 x 5 = 45
9 x 6 = 54

Explanation: In the result columns, write 0 to 9 from top to bottom in the ten's place as shown in blue and then write 0 to 9 from bottom to top in the units places. Writing 0 to 9 in reverse orders at two places (ten's place & unit's place) completes the 9 table. This is a pattern.

9 x 7 = 63
9 x 8 = 72
9 x 9 = 81
9 x 10 = 90

OBSERVATION:

(i) In the result section find that in all the rows the digits added together makes 9.
(ii) If the ten's place is written then the unit's place is the digit 9 minus the ten's place.
(iii) If the unit's place is written then the ten's place is the digit 9 minus the unit's place.

OBSERVATION:

Any number multiplied by 9, the total digit sum of the result will be 9 only.
Example: 9 x 12 = 108 and 1+0+8 = 9.
Example: 9 x 23 = 207 and 2+0+7 = 9.
Example: 9 x 62 = 558 and 5+5+8 = 18, 1+8 = 9.
Example: 9 x 132 = 1188 and 1+1+8+8 = 18, 1+8 = 9.

NOTE: We do any digit multiplication by 9, the digits sum of the product shall be 9 only. This is the beauty of multiplication by 9.

#Tr.4 **MULTIPLICATION BY 9:**

Multiplication of any number by 9 is too easy to conduct. After a little practice, you may not need any pen and paper and you can do the calculations in your mind itself.

METHOD 1:

Step 1: Multiply the given number by 10 and
Step 2: Subtract the given number from the above result.
Example: 15 x 9 = 15 x (10-1) = 150 − 15 = 135.

In the above example, we multiplied 15 by 10 & got 150. Next, we subtracted 15 from 150 & got 135
Example: 222 x 9 = 2220 − 222 = 1998.
Example: 121 x 9 = 1210 − 121 = 1089.
Example: 67 x 9 = 670 − 67 = 603.

#Tr.5 MULTIPLICATION BY 9:

METHOD 2:

Step 1: Make it into two parts putting a slash '/'.
Step 2: On the right part of the '/' write the complement from 10 of the unit's place digit of the given number.
Step 3: On the right part of '/' write the result obtained from (the given number − one more than rest of the number omitting unit's place)

Example: 25 x 9 = (25−(2+1))/$\overline{5}$ = 22/5 = 225.
Explanation:
Step 1: Slash is put in between.
Step 2: 5 is in the unit's place & whose complement is 5. We wrote 5 on the left part of the '/'.
Step 3: 2 is the number left with after omitting the unit's place & we calculate: 25 − 3 = 22.
Example: 121 x 9 = (121−13)/9 = 108/9 = 1089.
Explanation: 1's complement is 9 written on the left part. Omitting 1 in the unit's place, the number left is 12. So, 121 − 13 = 108.

Example: 83 x 9 = (83−(8+1))/$\overline{3}$ = 74/7 = 747.

Example: 74 x 9 = (74−(7+1))/6 = 66/6 = 666.
Explanation: 4's complement is 6. One more than 7 is 8 and 67 − 8 = 66.
Example: 237 x 9 = (237−24)/3 = 2143.

Dear learners, the author expects, you learned both of these methods in a delighted way and now is the time to practice doing some problems.

DO YOURSELF

1. Write the multiplications of the following numbers by 9: 12, 103, 333, 239, 780, 673, 543, 231, 117 & 707.

#Tr.6 MULTIPLICATION BY 11

The multiplication of any integer by 11 is very easy to perform. It can be performed as follows.

Method 1:
Step 1: Write zeros at the beginning and end of the given number. In fact, sandwich the given number between zeros.
Step 2: Starting from right end sum the digits to the left taking them in pairs.
Step 3: In case, the sum total exceeds 9, retain the first digit and carry over the other digits to the left.

Alternatively, to multiply a number of any digits, by 11 one can write down the number to be multiplied, and put the total of the digits between the two digits from the right to the left or vice versa.

NOTE: The unit place of the product is the unit place of the multiplicand.

Example: Multiply, 234, 451, 706, and 5432 by 11.

```
  234        351        706        5432
x  11      x  11      x  11      x  11
 2574       3861       7766      59752
```

Example: Multiply, 64, 457, 863, and 998 by 11.

```
   64        457        863        998
x  11      x  11      x  11      x  11
  704       5027       9493      10978
```

Explanation:

In 64 x 11, **4** is retained, 4+6=10, **0** is kept and 1 is carried to the left, added with 6, we got 7.**(704)**

In 457 x 11, **7** is retained, 5+7=12, **2** is kept and 1 is carried to the left, 4 added with 5, gives 9, 9 added with last carried 1, we get 10, where **0** is kept and 1 is carried to the left to be added with 4, gives **5.(5027)**

In 863 x 11, **3** is retained, 3+6=**9**, then 6 is added with 8, gives 14, **4** is kept and 1 is carried to the left, added with 8 gives **9.(9493)**

In 998 x 11, **8** is retained, 8+9=17, **7** is kept and 1 is carried to the left, 9 added with 9, gives 18, 18 is added with last carried 1, gives 19, where **9** is kept and 1 is carried to the left, added with 9, gives **10.(10978)**

NOTE: Alternatively, the same process can be done starting from left end reaching the right end to get the above results.

OBSERVATION: #DIVISIBILITY BY 11: In the products, we see that in all the examples, the sum of alternative digits (at even & odd places) are either equal (difference is zero) or differ by 11.

In 704, 7 + 4 = 11 is the sum of digits in odd places, and 0 is the even place. Their difference, 11 − 0 = 11.

In 5027, the sum of digits in odd places is 7 + 0 = 7 and that of the digits in the even places is 2 + 5 = 7. Their difference, 7 − 7 = 0.

In 9493, the sum of digits in odd places is 3 + 4 = 7 and that of the digits in the even places is 9 + 9 = 18. Their difference, 18 − 7 = 11.

In 10978, the sum of digits in odd places is 8 + 9 +1 = 18 and that of the digits in the even places is 7 + 0 = 7. Their difference, 18 − 7 = 11.

CONCLUSION: If the difference of "sum of the digits in even places and sum of the digits in odd places" is either zero or a multiple of 11, then the number is **divisible by 11**.

DO YOURSELF

1. Write the multiplications of the following numbers by 11: 27, 213, 555, 737, 780, 671, 999, 1099, & 7907.

#Tr.7 MULTIPLICATION BY 12

Method 1: Multiplying any number by 12 can be performed by splitting 12 into 10 & 2.

Example: 23 x 12 = 23 x 10 + 23 x 2 = 276
Example: 337 x 12 = 337 x 10 + 337 x 2 = 3370+774 = 4044.

Method 2: We will use the Vedic rule, called 'Ultimate and twice the penultimate'. Penultimate is the last but one.

Step 1: Sandwich the multiplicand between zeros

Step 2: Add twice the penultimate digit to the ultimate digit

Step 3: Work from right to left/from left to right.

NOTE: In case, the sum total exceeds 9, retain the first digit and carry over the other digits to the left.

Example: Find the value of 134 x 12.

01340   Penultimate is 4 working from the right
x   12   Add (2 x 4)+0=**8**
─────
1608   Add (2x3) +4=10, **0** is kept & 1 carried

Add (2x1) + 3=5, 5+1 carried is **6**

Add (2x0) +1=**1**   **OR**

01340   Penultimate is 1 working from the left
x   12   Add (2 x 0)+1=**1**
─────
1608   Add (2x1) +3=5, **6** is kept (5+1 carried)

Add (2x3) + 4= 10, 0 is kept & 1 carried

Add (2x4) +0=**8**

NOTE: It is suggested to practice from left to right as well.

Multiplication by 13: The multiplication of any number by 13 can be performed as follows.

Step 1: Append '0' after and before the number, to be multiplied by 13.

Step 2: Starting with the unit's place digit of the given number, multiply it by 3 & add the result to the next digit on the right. Repeat the procedure for each of the remaining digits.

NOTE: In case, the sum exceeds 10, carry over the ten's place digit to the left & retain the unit's place in the result. *We can also do the multiplication, directly without putting 0 at the ends of the number.*

Example: 123 x 13 (=1599)
= 01230 x 13 (0 is appended after & before)
Calculations:
3 x 3 + 0 = **9**(take 9 as the units place of the product)

3 x 2 + 3 = **9**(take 9 as the ten's place of the product)
3 x 1 + 2 = **5**(take 5 as hundred's place of the product)
3 x 0 + 1 = **1**(take 9 as thousand's place of the product)

Example: 567 x 13 (=7371)
= 05670 x 13 (0 is appended after & before)

Calculations:
3 x 7 + 0 = **21,** 2 is to be carried
3 x 6 + 7 = 25, 25 + 2 (carried)=**27,** 2 is to be carried
3 x 5 + 6 = 21, 21 + 2 (carried) =**23,** 2 is to be carried
3 x 0 + 5 =5,5+2 (carried) = **7**

#Tr.8 BASE METHODS

REGULAR METHOD: We have been taught starting from our school days, to perform the multiplication in the following way.

Example: Solve 12 x 13

Consider   12
            x 13
             36   (12 multiplied by 3, unit place of 13)
           12____ (12 multiplied by 1, ten's place of 13)
         156

(we added both the rows, in columns)

We use the concept of base to ease the multiplication of numbers near to such bases. Consider the bases as 10, 100, 1000, 10000, etc. Let us learn the tricks of multiplications in noting down how far away are these numbers from their respective bases.

NEAR TO THE BASE 10:

CASE I: When multiplication of unit place digits is less than 10.

Consider 12 (+2), 12 is more than 10 by 2

        x13 (+3), 13 is more than 10 by 3
            15/6 (2 x 3 = 6 and 12 +3 = 15 or 13 + 2 = 15)
            156

Explanation:
Step 1:
We write +2, as 12 is more than 10 by 2 (excess).
We write +3, as 13 is more than 10 by 3 (excess).
Step 2:
In the result, make it in to two parts and separate them by a '/'
Write theproduct of 2 and 3 on the left part of '/'. In the right part of '/' write the addition of one number with the excess (or less) of another number from the base, 10. Here it is either 12 + 3 =15 or 13 + 2 = 15.
Step 3:
Remove the '/' sign. Here the result is 156.

CASE II: When multiplication of unit place digits is more than 10.
Example: Multiply 13 by 14
Consider   13 (+3), 13 is more than 10 by 3
            x 14 (+4), 14 more than 10 by 4
            17/$_1$2 (3 x 4 = 12 & 13 +4 = 17 or 14 + 3 = 17)
   182

NOTE: Here 3 x 4 = 12. 2 is kept & 1 is carried to left.

Example: Multiply 23 by 13
Consider   23 (+13), 13 is more than 10 by 13
            x 14 (+4), 14 more than 10 by 4
            27/$_5$2 (13 x 4 = 52 & 23 +4 = 27 or 14+13= 27)
            322

NOTE: Here 13 x 4 = 52. We keep 2 in the unit's place and carry '5' to the right and add it to 27.

CASE II: When one of the numbers is less than 10.
Example: Multiply 17 by 8
Consider   17 (+7), 17 is more than 10 by 7

Fifty Tricks of Multiplication

$$\begin{array}{r} \text{x } 8 \;(-2), \text{ 8 is less than 10 by 2} \\ \hline 15/\overline{1}4 \,(7 \times (-2) = -14 = \overline{1}4 \;\&\; 17-2=15 \text{ or } 8+7=15) \\ \hline 136 \quad (150 - 14 = 136) \end{array}$$

NOTE: Here $7 \times (-2) = -14 = \overline{1}4$.
Now $15/\overline{1}4 = 150 - 14 = 136$ (15 in the ten's place & has the value 150)

Example: Multiply 24 by 9
Consider   23 (+13), 24 is more than 10 by 14
$$\begin{array}{r} \text{x 9 } (-1), \text{ 9 is less than 10 by 1} \\ \hline 23/\overline{1}3 \,(13 \times -1) = -13 = \overline{1}3 \;\&\; 23-2=22 \text{ or } 9+13=22) \\ \hline 217 \quad (230 - 13 = 217) \end{array}$$

#NEAR TO THE BASE 100:

Here we will multiply the numbers near to 100. These numbers may be more than or less than 100. Each will be explained as the case may be.

Consider 107 (+07), 107 is more than 100 by 7
$$\begin{array}{r} \text{x 113 } (+13), \text{ 113 is more than 100 by 13} \\ \hline 120/91 \; (13 \times 7 = 91 \text{ and } 107+13=120 \\ 12091 \; \text{ or } 113 + 7 = 120) \end{array}$$

NOTE: Here 100 has two zeros, and on the left part of the '/' we keep two digits.
In case it is a single digit we write a zero on its right and write it. If the product is a 3 digit number we keep 1st two digits and carry the 3rd digit to the left for addition.

Example: Multiply 112 by 113
Consider 112 (+12), 112 is more than 100 by 12
$$\begin{array}{r} \text{x113 } (+13), \text{ 113 is more than 100 by 13} \\ \hline 125/_156 \,(13 \times 12=156 \;\&\; 112+13=125 \text{ or } 113+12=125) \\ \hline 12656 \end{array}$$

NOTE: Here $13 \times 12 = 156$ has 3 digits. We kept 56 on the right part and carried 1 to the left part for addition (125+1=126).

Example: Multiply 108 by 93

Consider 108 (+8), 108 is more than 100 by 8
$\quad$ x 93 (−7), $\quad$ 93 is less than 100 by 7
$\overline{101/\overline{5}6}$(8x(-7)=-56=$\overline{5}$6 & 108-7=101 or 93+8=101)
$\quad\;\;\overline{10044}$

NOTE: Here 8 x (−7)= −56=$\overline{5}$6
In 101$^{\overline{5}6}$, 101 is in hundred's place and so,
101$^{\overline{5}6}$= 10100 − 56 = 10044.

Example: Multiply 98 by 96
Consider 98 (−2), 98 is less than 100 by 2
x93 (−7), 93 is less than 100 by 7
$\quad\;\;\overline{91/14}$ ((−2)x(−7)=14 & 98 −7=91 or 93 −2=91)
$\quad\;\;\;\overline{9114}$

SUMMARY:

I. Near to the base, for more we use '+' sign and for less, we use '−' sign.

II. Based on the number of zeros in the base, we keep the same number of digits on the left part of the '/'sign, and carry the rest digits to the right and add, as the case may be.

III. Similar procedure can be followed for the multiplication of numbers near to the base 10000 etc.

PROPOSAL: We know practice makes perfect. It is expected that you finish up the following multiplications in mind and write down the answers.

DO YOURSELF:

1. Solve 24 x 13, 9 x 19, 8 x 17, 27 x12 near the base 10.

2. Solve 112 x 98, 106 x 99, 92 x 91 near the base 100.

#Tr.9 SPLIT & MERGE METHOD

In this method, we split one of the numbers (multiplicand / multiplier or both), perform the small multiplications and then add the results.

Example: Find the value of 23 x 12.

Case I: Here we split 12 as (10+2).

23 x 12 = 23 x 10 + 23 x 2 = 230 + 46 = 276

Case II: Here we split 23 as (20+3).

23 x 12 = 12 x 20 + 12 x 3 = 240 + 36 = 276

Case III: Here we split 23 as (13+10).

23 x 12 = 13 x 12 + 10 x 12 = 156 + 120 = 276

Case IV: Here we split 23 as (13+10).

23 x 12 = 13 x 10+10 x 12+13 x 2 = 130+120+56 = 276

NOTE: It is the smartness of our mind & it explores the right split and complete the multiplication. Perhaps, Case I is smarter in comparison.

Example: Find the value of 23 x 12.

Case V can be 23 = 20 + 3 and 12 = 10 + 2 and the calculation in the mind can be:

200(20x10) + 40(20x2) + 30(10x3) + 6(3x2) = 276.

So, it is a business of our mind to decide, which is the quicker one.

Example: Find the value of 713 x 8.

We can split 713 as 700 + 13 and so the product is (700 x 8) + (13 x 8) = 5600 + 104 = 5700.

Fifty Tricks of Multiplication

NOTE: Doing all these calculating in your mind is desired in the process of practicing. You need to do all in your mind only.

DO YOURSELF:

1. Solve 14 x 13, 99 x 19, and 18 x 17 near the base 10.

2. Solve 112 x 98, 106 x 99, and 92 x 91 near the base 100.

#Tr.10 MULTIPLY TEEN NUMBERS

Teen numbers are said to be between 10 and 20. We multiply couple of teen numbers directly as follows.

Step 1: Add the unit's place digits of both the numbers.
Step 2: Add 10 to the result obtained in step 1.
Step 3: Multiply the result of step 2 by 10
Step 4: Multiply the unit's place digits of both the numbers.
Step 5: Add the results of Step 3 and step 4.

Example: 16 x 18=(14+10)x10+6x 8=240+48=288
Explanation:
The unit's place digits are 6 & 8. Added together gives 14. Now, 14 + 10 = 24. 24 multiplied by 10 is 240. Next, the product of 6 & 8 is 48. So 240 + 48 = 288.

Example: 14 x 17=(11+10)x10 +4 x 7=210+28=238

Example: 13 x 19= 247.

Dear readers, now try to do it in your mind directly. Yes, you can make it within seconds.

DO YOURSELF:

1. Try the following multiplications of teen numbers: 14 x 19, 13 x 18, 17 x 16, 12 x 18, 14 x 16 and 19 x 16.

#TR.11 MULTIPLICATION OF 3 DIGIT NUMBERS WITH SAME HUNDRED'S PLACE:

Step 1: Add the ten's place digits of both the numbers. Multiply it by "the hundred's place digit"
Step 2: Add 100/ 200, as the case may be to the result obtained in step 1.
Step 3: Multiply the result of step 2 by 100/200 as the case may be
Step 4: Multiply the ten's place digits of both the numbers.
Step 5: Add the results of Step 3 and step 4
Example: Evaluate 123 x 112. Hundred's place is 1.
23 + 12 = 35. 35 + 100 = 135. 135 x 100 = 13500
23 x 12 = 276. The answer is 13500 + 276= 13776.
Example: Evaluate 213 x 224. Hundred's place is 2.
13 + 24 = 37. 37 + 200 = 237. 237 x 200 = 47400
13 x 24 = 312. The answer is 47400 + 312= 47712.
Example: Evaluate 333 x 312. Hundred's place is 3.
33 + 12 = 45. 45 + 300 = 345. 345 x 300 = 103500
33 x 12 = 396. The answer is 103500 + 396= 103896.

DO YOURSELF

1. Evaluate, 134 x 123, 243 x 224, 431 x 412, 512 x 543, 823 x 811, 646 x 622 and 732 x 743

CHAPTER THREE

In this chapter, we will basically learn the multiplication of two digit / three digit numbers in particular forms in pairs. During this learnings, we will use the Vedic sutra, 'Ekadhikena Purvena' means, 'one more than the previous one'. We will conduct the multiplication of numbers which are equidistant from their respective, bases. We will also use the Nikhilam / complement rule, in cases.

#Tr.12 NUMBERS ENDING WITH 1:

Here, we multiply two digit / three digit numbers ending with 1 (unit places are 1).
It involves 3 steps / 3 part calculations.

Step 1: Keep 1 in the unit's place (unit's places multiplied)

Step 2: Add the a and b

Step 3: Multiply a and b

$$\begin{array}{r} a1 \\ \times\, b1 \\ \hline ab/(a+b)/1 \end{array}$$

NOTE: In case, the sum total exceeds 9, retain the first digit and carry over the other digits to the left.

Example: Multiply 51 by 31, 61 by 91 and 131 by 131.

$$\begin{array}{r} 51 \\ \times\, 31 \\ \hline (5\times3)/(5+3)/1 \\ 1581 \end{array} \qquad \begin{array}{r} 61 \\ \times\, 91 \\ \hline (6\times9)/(6+9)/1 \\ 5551 \end{array}$$

Explanation:

In 51 x 31, 1 is kept in the unit's place, 5 is added with 3 gave 8, 5 multiplied by 3 gave 15. **(1581)**

In 61 x 91, 1 is kept in the unit's place, 6 is added with 9 gave 15, 5 is kept and 1 is carried to the left. Next, 6 is multiplied by 9 gave 54, added with last carried 1, we get 55. **(5551)**

In 131 x 131, 1 is kept in the Unit's place. 13 added with 13 gives 26; 6 is kept and 2 is carried to the left. Next, 13 is multiplied by 13, gives 169 added with 2 carried, we get 171. **(17161)**

$$\begin{array}{r} 131 \\ \times\ 131 \\ \hline (13 \times 13)/(13+13)/1 \\ 17161 \end{array}$$

It is expected that, now we are ready to do the calculations in our minds. Hopefully, you enjoyed this and keep it up.

DO YOURSELF:

1. Solve 31 x 91, 71 x 61, 141 x 121, 191 x 191, 211 x 221, 181 x 181, 161 x 131 using the above trick.

#Tr.13 NUMBERS WITH SAME TEN'S PLACES & UNIT'S PLACES ADD TO 10

Here, we multiply two numbers, whose unit's place digits adds to 10 and the ten's place digits are same. In this method, we shall use the Vedic sutra, called 'Ekadhikena Purvena' means, 'one more than the previous one'. In this trick, we need to multiply the ten's place by one more than the ten's place.

Step 1: Multiply the digits in the Unit's places of both the numbers.
Step 2: Multiply the ten's place digit by one more than the ten's place digit.

Fifty Tricks of Multiplication

Example: Find the values of 64 x 66, 73 x 77, 82 x 88 and 91 x 99.
64 -Both the numbers have same ten's place, 6
x **66** - The unit's places 4 & 6 adds to 10
$\overline{4224}$ - [4 x 6 = 24, 6 x (6+1) = 42]

73 - Both the numbers have same ten's place, 7
x **77** - The unit's places 3 & 7 adds to 10
$\overline{5621}$ - [3 x 7 = 21, 7 x (7+1) = 56]

82 - Both the numbers have same ten's place, 8
x **88** - The unit's places 2 & 8 adds to 10
$\overline{7216}$ - [2 x 8 = 16, 8 x (8+1) = 72]

91 - Both the numbers have same ten's place, 9
x **99** - The unit's places 1 & 9 adds to 10
$\overline{9009}$ - [1 x 9 = 09, 9 x (9+1) = 90]

NOTE: In the above example, 1 x 9 = 9, but we write it 09, to make it two places.

DO YOURSELF:

1. Solve 13 x 17, 71 x 79, 48 x 42, 96 x 94, 22 x 28, 97 x 93, 67 x 63 using the above trick. Of course, it is expected you solve these within your mind.

#Tr.14 NUMBERS WITH SAME UNIT'S PLACES & TEN'S PLACES ADD TO 10

Here, we multiply two numbers, whose unit's places are same and the sum of the ten's place digits is 10. We solve these multiplication using 2 easy steps.

Step 1: Multiply the digits in the Unit's places of both the numbers.
Step 2: Multiply the ten's place digits and the product with the common unit place digit.

Fifty Tricks of Multiplication

Example: Find the values of 46 x 66, 37 x 77, and 28 x 88.

46 -Both the numbers have same unit's place, 6
x66 - The ten's places 4 & 6 adds to 10
―――
3036

Explanation:
- Unit's place digits are multiplied, 6x6=36,
- Ten's place digits, 4 & 6 are multiplied, 4 x 6=24, and then added with the common unit's place digit, 6, so we get 24 + 6 = 30]. (**3036**)

37 -Both the numbers have same unit's place, 7
x 77 - The ten's places 3 & 7 adds to 10
―――
2849

Explanation:
- Unit's place digits are multiplied, 7 x 7=49,
- Ten's place digits, 3 & 7 are multiplied, 3x7=21, and then added with the common unit's place digit, 7, so we get, 21 + 7 = 28]. (**2849**)

28 -Both the numbers have same unit's place, 8
x 88 - The ten's places 2 & 8 adds to 10
―――
2464

Explanation:
- Unit's place digits are multiplied, 8 x 8=64,
- Ten's place digits, 2 & 8 are multiplied, 2x8=16, and then added with the common unit's place digit, 8, so we get, 16 + 8 = 24]. (**2464**)

NOTE: If the multiplication of the unit's place digits is a single digit, put a zero on the right of it.

DO YOURSELF:

1. Solve 33 x 73, 86 x 26, 41 x 61, 96 x 16, 28 x 88, 84 x 24, 66 x 46 using the above trick. Never use pen and paper. But do it in your mind.

#Tr.15 NUMBERS WITH SAME TEN'S PLACES & SUM OF THE UNIT'S PLACES IS MORE THAN 10:

Here, we multiply two numbers, which has same ten's places but the sum of the Digits in the unit's places is more than 10. It is the time to remember that, the Vedic sutra 'Ekadhikena Purvena' / 'One more than the previous one' applies, whenever the ten's places are same.

Step 1: Multiply the unit's place digits. Write it on the right part of '/'.
Step 2: On the left side of '/' write (ten's place digit) x (ten's place digit+1)
Step 3: Then add ten's place value multiplied by the sum of unit's places, excess of 10.

Example:

  78 -Both the numbers have same ten's place, 7
x 76 - The unit's places, 8+6=14, **4** more than 10
56/48- Write 7 x (7+1)=56, 7 is the ten's place
+280 -Add 4 x 70 = 280 (Ten's place value is 70)
5928   (Since it is 4 more & ten' place is 7)

Example:

  86 -Both the numbers have same ten's place, 8
x 87 - The unit's places, 6+7=13, **3** more than 10
72/42- Write 8 x (8+1)=72, 8 is the ten's place
+240 -Add 3 x 80 = 240(10's place value is 80)
7482   (Since it is 3 more & ten' place is 8)

Example:

  56 - Both the numbers have same ten's place, 5
x 59 - The unit's places, 6+9=15, **5** more than 10
30/54 - Write 5 x (5+1)=30, 5 is the ten's place
+250 - Add 5 x 50 = 250 (Ten's place value is 50)
3304   (Since it is 5 more & ten' place is 5)

#Tr.16 NUMBERS WITH SAME TEN'S PLACES & SUM OF THE UNIT'S PLACES IS LESS THAN 10:

Here, we multiply two numbers, which has same ten's places but the sum of the Digits in the unit's places is less than 10.

Step 1: Multiply the unit's place digits. Write it on the right part of '/'.
Step 2: On the left side of '/' write (ten's place digit) x (ten's place digit+1)
Step 3: Then subtract ten's place value multiplied by the sum of unit's places, less of 10.

Example:

  74 - Both the numbers have same ten's place, 7
x 73 - The unit's places, 4+3=7, **3** less than 10
56/12 - Write 7 x (7+1)=56, 7 is the ten's place
−210 - Subtract 3 x 70 = 210 (Ten's place value is 70)
5402   (Since it is 3 less & ten' place is 7)

Example:

  81 - Both the numbers have same ten's place, 8
x **85** - The unit's places, 1+5=6, **4** less than 10
72/05 - Write 8 x (8+1)=72, 8 is the ten's place
+320 - Add 4 x 80 = 320 (Ten's place value is 80)
6885   (Since it is 3 more & ten' place is 8)

Example:

  92 -Both the numbers have same ten's place, 9
x **9**6 - The unit's places, 2+6=8, **2**less than 10
90/12 - Write 9 x (9+1)=90, 9 is the ten's place
−180 -Add 2 x 90 = 180 (Ten's place value is 90)
8832 (Since it is 5 more & ten' place is 5)

DO YOURSELF:

1.Find the values of 33 x 39, 66 x 68, 48 x 44, 91 x 99, 78 x 79 and64 x 69 using the above trick. Try to do it in your mind.

2.Find the values of 46 x 41, 66 x 62, 81 x 84, 19 x 11, 85 x 81 and77 x 71 using the above trick. Try to do it in your mind.

#Tr.17MULTIPLICATION OF TWO CONSECUTIVE TWO DIGIT NUMBERS:

We will do it by splitting the multiplicand from the multiplier. Let 'a' is the multiplicand and (a+1) is the multiplier, so we get a (a+1) = a^2 +1.

Example:
12 x 13 = 12 x (12 + 1) = 12 x 12 + 12 = 12^2 + 12 =156

Example:
17 x 18 = 17 x (17 + 1) = 17 x 17 + 17 = 17^2 + 17 =306

Example:
21 x 22 = 21^2 + 21 =441 +21 = 462.

Example:
55 x 56 = 55^2 + 55 = 3025 +55 = 3080.

DO YOURSELF:
1.Try these: 23 x 24, 95 x 96, 41 x 42 & 88 x 89 now.

#Tr.18 MULTIPLICATION BY 50:

When any number is to be multiplied by 50, you need to half it once and then multiply it by 100, to get the result.

$Motivation: 50 = \dfrac{100}{2}$, so $a \times 50 = a \times \dfrac{100}{2} = \dfrac{a}{2} \times 100$

Example: 36 x 50 = 18 x 100 = 1800
Explanation: 36 is halved & multiplied then by 100.
Example: 135 x 50 = 62.5 x 100 = 6250
Explanation: 135 is halved & multiplied by 100.
Example: 775 x 50 = 387.5 x 100 = 38750
Example: 97 x 50 = 48.5 x 100 = 4850
Example: 9950 x 50 = 4975 x 100 = 49750

DO YOURSELF:

1. Do multiply 23, 95, 341, 772, 980, 9898, 1234 & 8815, 788, 342 by 50.

#Tr.19 MULTIPLICATION OF NUMBERS EQUIDISTANT FROM A DECLARED BASE:

Let A be the base. Then A + d and A − d are two numbers equidistant from A, with distance d. Knowing the base and distance the multiplication is (A+d) x (A−d) = $A^2 - d^2$.

Example: Find 56 x 44.
Here, base = (56+44)/2 = 50 = A
And d = 6.
Now, 56 x 44 = (50+6) x (50−4) = $50^2 - 6^2$ = 2464.

Explanation: Here A = 50 is taken as base and it is equidistant from 56 & 44 by a distance 6.

Example: Find 17 x 23.
Here, base = (17+23)/2 =20 =A
And d = 3.
Now, 17 x 23 = (20+3) x (20−3)
= $20^2 - 3^2$ = 391.

Example: Find 58 x 62.
58 x 62 = $60^2 - 2^2$ = 3600 − 4 = 3596. (A=60, d = 2)

Example: Find 58 x 62.
91 x 109 = $100^2 - 9^2$ = 10000 − 81 = 9919. (A=100, d=9)

DO YOURSELF:

1. Do multiply 23 x 17, 95 x 105, 34 x 26, 72 x 78, 980 x 1020, 89 x 111, and 76 x 84.

#Tr.20 MULTIPLICATION OF NUMBERS WITH DISTANCE 1 FROM A NUMBER:

It is a particular case of the above case, where d is 1.
So $A^2 - d^2$ becomes $A^2 - 1$.

Example: 17 x 19 = $18^2 - 1$ = 324 − 1 = 323

Explanation: 17 and 19 are at a distance of 1 from 18. So $18^2 - 1^2$ = 324 − 1 = 323.

Example: 23 x 25 = $24^2 - 1$ = 576 − 1 = 575
Example: 99 x 101 = $100^2 - 1$ = 9999.

Example: 109 x 111 = $110^2 - 1$ = 12100 − 1 = 12099.
Example: 74 x 76 = $75^2 - 1$ = 5625 − 1 = 5624.

Example: 26 x 28 = $27^2 - 1$ = 729 − 1 = 728.

Fifty Tricks of Multiplication

DO YOURSELF:

1. Find the values of 23 x 17, 95 x 85, 341 x 259, 72 x 68, 121 x 79, 78 x 62, and 42 x 58.

2. Find the values of 101 x 99, 79 x 81, 23 x 25, 34 x 36, 89 x 91, 121 x 119 and 54 x 56.

#Tr.21 MULTIPLICATION OF NUMBERS WITH 5 IN THE UNITS PLACE:

If the unit's places are 5 in both the numbers and

Category I: the ten's places add to 10, then follow:
65 x 45 = ((6x4)+5) / (5x5) = 2925
95 x 15 = ((9x1)+5) / (5x5) = 1425
75 x 35 = ((7x3)+5) / (5x5) = 2625

Explanation: Unit places are multiplied and kept on the right part of '/'. Ten's places are multiplied and added with 5 in each case for the left part of the product.

Category II: the ten's places do not add to 10, then we proceed as follows:
Let the numbers be a5 x b5
= (10a+5) x (10b+5)
= 100 ab + 50(1+b) + 25
= 100 (ab + (a+b)/2) + 25

Example: 85 x 45 = 100 ((8x4) + (8+4)/2) + 25
= 3800 + 25 = 3825

Example: 95 x 35 = 100 ((9x3) + (9+3)/2) + 25
= 3300 + 25 = 3325.

NOTE: When a and b both are even or both are odd, in the result we end up with 25.

Example: 95 x 65 = 100 ((9x6) + (9+6)/2) + 25

Fifty Tricks of Multiplication

= 100 (54 + 7.5) + 25
= 100 (61.5) + 25
= 6150 + 25 = 6175.
Example: 115 x 125 = 100 ((11x12) + (11+12)/2) + 25
= 100 (132 + 11.5) + 25
= 100 (143.5) + 25
= 14350 + 25 = 14375.

NOTE: When a and b are not both even / both odd, we end up with 75 in the result.

Do it in your mind:

Example: 85 x 65 = 55/25=5525? ((8 x 6) +(8+6)/2)
Example: 115 x 95 = 10925?((11 x 9) +(11+9)/2)
Example: 135 x 55 = 7425?
Example: 75 x 95 = 7125?
Example: 15 x 25 = 375?

Explanation: Here 1 x 2 = **2** and (1+2)/2 = **1**.5. In this case the reader can directly do it like 375 (Add 2 with 1 ignoring .5). And then write 75 in the first two places.

Example: 55x85 =4675?

(8x5)=40 and (5+8)/2=6.5. We add 40 with 6 and get 46, ignoring .5 & write 75 in the first two places. Now the final product is 4675.

Example: 115 x 65 = 7475? **((11 x 6) +8)**

SUMMARY: If the unit's places is 5, tens places are both even or both odd, we get the product ends up with 25 and else it ends up with 75.

SUGGESTION: Solve the following in your mind.

Fifty Tricks of Multiplication

DO YOURSELF:

1. Find the values of 25 x 85, 95 x 35, 115 x 135, 75 x 95, 125 x 105, 85 x 65, and 55 x 105.

2. Find the values of 105 x 95, 75 x 88, 25 x 75, 35 x 65, 185 x 55, 125 x 115 and 65 x 55.

CHAPTER FOUR

In this chapter we will learn some interesting tricks of multiplications that uses graphical images and or geometrical figures. Across the countries, people uses different tools for doing the multiplications.

#Tr.22 HALVE& DOUBLE METHOD/ RUSSIAN PEASANT METHOD:

This is a trick, which uses the process of halving one number and doubling the other and some other steps. Make two columns and write one number in each column. Refer the following steps to find the product.

Step 1: For any two numbers, halve the 1st number repeatedly, discarding any fractional remainder, until you reach at 1.

Step 2: Simultaneously, double the 2nd number as many times as you halved the first.

Step 3: Cross out every even number in the halves column and the corresponding number in the doubles column.

Step 4: Add the numbers that remain in the doubles column only, & the sum is the product of the two numbers.

Example: Find the value of 122 x 231

Halves	Doubles	
~~122~~	~~231~~	
61	462	
~~30~~	~~924~~	
15	1848	
7	3696	
3	7392	
1	14784	
	28182	

Steps are followed as stated. We stroked out even numbers in the halves column & corresponding in the doubles column. We add the rest in the doubles column is the product of the given numbers.

Fifty Tricks of Multiplication

Example: Find the value of 28 x 16, 24 x 41.

Halves	Doubles	Halves	Doubles
~~28~~	~~16~~	~~24~~	~~41~~
~~14~~	~~32~~	~~12~~	~~82~~
7	64	~~6~~	~~164~~
3	128	3	328
1	256	1	656
	448		984

DO YOURSELF

1. Use double and halve method to find the values of 23 x 44, 17 x 77, 91 x 122, 43 x 98 and 102 x 113.

#Tr.23 CROSS LINES/STICK METHOD

It is a method practiced in schools in China, Japan, Korea. It uses a set of crossed lines representing the couple of numbers to be multiplied. The number of dots (intersections) of these crossed lines are counted and arranged to find the product. The steps of this method is as follows:

Step 1: Draw number of lines = Unit place digit of one number from left to right inclined, then with a gap draw number of parallel lines = ten's digit place and repeat for the rest of the digits. (As shown in the figure)

Step 2: Draw the number of lines crossing the previous set of lines following the procedure in Step 1, but opposite in direction, as shown in figure.

Step 3: Count the number of points of intersections at each group and write them in the same order as that of the groups of points of intersections, is the required result of the product.

NOTE: In case the number of points of intersections exceeds 9, carry the rest digits to the left for addition.

Fifty Tricks of Multiplication

Example. Find the value of 22 x 32
Draw 4 parallel lines to represent 22, the first number in the product. Draw 2 lines, and then, a little further to the right, draw 3 more lines. These lines represent 22. Similarly draw 5 more lines to cross the previous 4, 3 lines to the left and 2 lines on the right. These will represent the number 32, the 2nd number in the product. Now we will count how many times all of the lines intersects.
We see that there are 4, 10 and 6 intersections. Combining from left to right we get 704 (from 10, we kept 0 and carried 1 to the left. So, **22 x 32 = 704.**

Example: Evaluate 124 x 14
We draw 7 parallel lines to represent the 1st number 124. Draw 1 line and then 2 more Lines to the right, and then 4 more lines to the right. Then drawing the lines in the opposite direction, to represent the 2nd number 14 with one line, and then 4 more lines to the right.

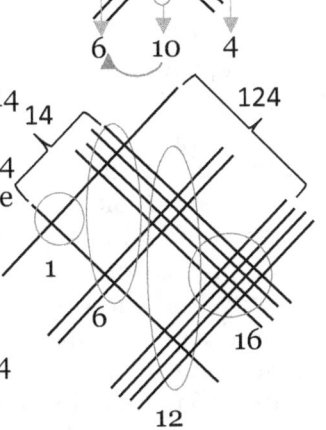

The number of intersects are 1, 6, 12 and 16 and 1 is to be carried from 16 and 12 to the group to their left groups. The final product shall be 124 x 14 = 1736
Note: Suppose a number in the multiplication includes '0', we put a dotted line and does not count the intersections generated from this line.

Fifty Tricks of Multiplication

Example: 203 x 21.
Draw lines as above to
Represent 203 and 21.
Instead draw a dotted line
to represent '0'.
Count the number of
Intersects, and do not
Include the intersects
Generated by the dotted line.
So there are 4, 2, 6 and 3
intersections, by combining
the numbers from the left to
the right, we get 4263.

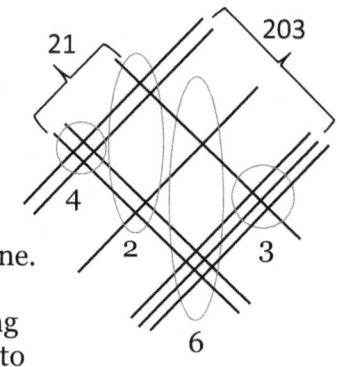

DO YOURSELF

1. Use cross lines / stick method to evaluate: 23 x 44, 17 x 77, 91 x 122, 43 x 98 and 102 x 113.

#Tr.24 ONE FIGURE & 4 PRODUCTS:

Suppose, we interpret the above method, in the way by rotating it by an angle 900, we will be able to see that it gives the answers to four different multiplication problems.
Let us have a look into this.

Example: Evaluate 42 x 21.
In the process of evaluation, we will see that, it will not only find the product value of 42 x 21 but also the product values of 24 x 21, 24 x 12 and 42 x 12.
Consider 42 x 21.
We draw 6 parallel lines to
represent the 1st number 42
Draw 4 lines and then 2 more
lines to the right. Then drawing
the lines in the opposite direction,
to represent the 2nd number 21
with 2 lines, and then 4 more lines to the right.

Fifty Tricks of Multiplication

So there are 8, 8, and 2 intersections. By combining the numbers from the left to the right, we get **882**.
Rotating the above figure by an angle of 90°, we get a different problem: 24 x 21 and the product is **504**.
{the intersections are 4, 10 & 4, And combining from left to right , we get 504, where 1 is carried from 10 to the left group}
Further rotation of 90°, generates a new problem, 24x12. And the product value shall be 288. {the intersections are 2, 8 and 8 and combining together from left to right, we get 288}
Further rotation of 90°, generates another new problem, 12 x 42 and the product value is 504

Note: Here we observed that the products 12 x 42 and 24x21 have the same value 504.

DO YOURSELF
1. Use cross lines / stick method to evaluate: 23 x 44, 17 x 77, and 91 x 12 to generate new problems.

#Tr.25 **SLIDING METHOD:**
Let us learn this method of sliding directly through an example.
Step 1: Write the reverse of one of the numbers given.
Step 2: Write the 2nd number below the number you reversed, keeping the unit place of the 2nd number

just below the last digit place of the reversed number.

Step 3: Write the product of these two digits in the column.

Step 4: Slide the number to the right. Multiply digits in the respective columns and add them. Write the result.

Step 5: Slide again to the right and repeat the procedure of step 4, until the digits exhausted.

Example: Evaluate 23 x 12.
Reverse 23 to get 32.

Write **32**   **32**   32
   12   12   12
   ―――   ―――   ―――
   6(3x2) **7** (3x1+2x2)   **2** (2x1)

The product is 276.

Or reverse 12 to get 21
Write **21**   21   21
   23   23   23
   ―――   ―――   ―――
   6(2x3) **7**(2x2+1x3) **2**(1x2)

The product is 276.

Example: Evaluate 321 x 211.
Reverse 321 to get 123.

  12**3**   1**23**   **123**   **123**   123
 211   211   211   211   211
 ―――   ―――   ―――   ―――――   ―――
 1   3   7   7(2x2+3x1)   **6**(3x2)
(1x1)   (1x1+2x1) (1x2+2x1+3x1)

The product is 67731.

You can also try taking the reverse of 211 & proceed, and get the same result.

NOTE: In case the sum of the products of columns exceeds 9, the rest digits to be carried to the left.

DO YOURSELF

1. Do practice some sliding problems as explained above for the following: 324 x 231 and 451 x 232.

Fifty Tricks of Multiplication

#Tr.26 LATTICE MULTIPLICATION:

Lattice multiplications, also known as Italian method, Chinese method, sieve multiplication, diagonally or Venetian squares, is a method of multiplication that uses a lattice to multiply two multi digit numbers.

Briefly, we can say, a simple and effective way to do any complex multiplication is lattice multiplication.

The simple procedures of lattice multiplication is stated in the following steps:

Step 1: Draw a table with m x n number of columns and rows, where m corresponds to the number of digits in the multiplicand and n – the number of digits in the multiplier.

Step 2: Put the digits of Themultiplicand aligned with the columns on the top of the table. Similarly, align the digits of the multiplier with the rows & place them on the right side of the table.

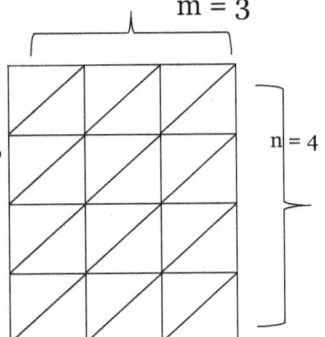

Step 3: Draw diagonals in each cell of the lattice from upper right to lower left to bisect each cell.

Step 4: Calculate a product for each cell by multiplying the digit at the top of the column and the digit at the right of the row. The tens digit is placed above the diagonal that passes through the cell and the units digit is put below the diagonal. If the product is less then 10, we put a zero above the diagonal in the corresponding cell.

Step 5: Start adding the numbers on the same diagonal path. Place the results on the left side of the diagonal paths. If the addition value exceeds 10, then keep the units digit and carry the rest digits to the left for additions to the corresponding value.

Step 6: Combine the digits, from top to bottom and then from left to right.

Example: Evaluate 234 x 36.
The answer is: 8424.
Explanation: We draw a table with three Columns (234 has 3 digits) & two rows (36 has 2 digits).
Cells are divided by equally

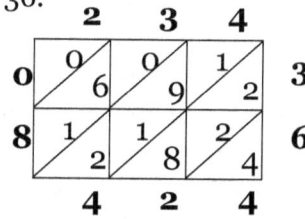

by diagonals. We multiplied the digits on the top column with the right side digits of the table, and placed the product in the cell partitions.

In the addition process of the diagonal path digits, 4 is placed as it is being alone, next, 8+2+2 = 12, we kept 2 and carried 1 to the left. In the 3rd path, 2+1+9+1= 13 and 1 was carried from previous path, so it will be 13 = 1 = 14, we kept 4 and carried 1 to the left. Next it is 1+6 = 7 & 7 added with 1, gives 8. The last path is with 0. Combining all the digits outside the table from top to bottom and then from left to right we get: 08424, that is **8424**.

Example: Evaluate 578 x 823.

We draw a 3 x 3 table with 3 columns and 3 rows and partition each cell equally by the diagonals.

578 is placed on the top of the columns and 823 on the right side of the table, aligned with the rows of the table.

Column by row digits multiplications are carried out and the results are placed in the upper and lower parts of the cell diagonals. After diagonal path additions obeying the rules, we got **475,694.**

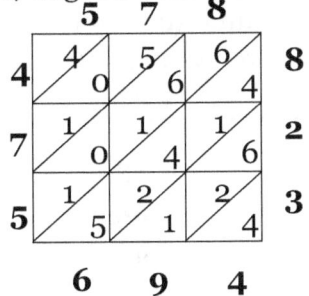

DO YOURSELF

1. Evaluate 323 x 33, 513 x 142 & 736 x 31 using the lattice multiplication method.

#Tr.27 EGYPTIAN METHOD:

This method never needed the multiplication table, but one needs to divide and multiply by 2 as follows:

Step 1: It uses two columns. Write the two numbers at the top of the columns.

Step 2: Begin with the number 1 and then keep on doubling it in the same column, like 2, 4, 8, 16, ... and stop only when you feel the next doubling will be more than the given number on the top.

Step 3: In the 2nd column, begin with the other number and keep on doubling aligning the rows in the 1st column, unless all is completed.

Step 4: Find the combination of numbers in the 1st / left column which adds to the 1st number.

Step 5: Ignore all the rows, in the table above if the

number is not included in the above sum process.

Step 6: Add up the remaining numbers in the 2nd column is the value of the product.

Example: Evaluate 142 x 24

142	**x**	**24**
1		24
2		**48**
4		**96**
8		**192**
16		384
32		768
64		1536
128*		**3072**

Explanation:

*We stopped at 128, since 128 doubles to 258 > 142.
We see that 128 + 8 + 4 + 2 = 142.

Ignore all other rows and adding the corresponding values in the rows of the 2nd column, we get:

3072 + 192 + 96 + 48 = 3408

NOTE: It comes to our mind, that, suppose we take 24 as the 1st number and 142 as the 2nd, what will happen? Let us try it.

24	**x**	**142**
1		142
2		284
4		568
8		**1136**
16*		**2272**

Explanation:

*We stopped at 16, since 16 doubles to 32 > 24.
We see that 16 + 8 = 24.

Ignore all other rows and adding the corresponding values in the rows of the 2nd column, we get:
2272 + 1136 = 3408.

OBSERVATION: We get the same answer, irrespective of selection of 1st number and 2nd number.

#Tr.28 MULTIPLICATION BY, 99, 999:

The multiplication of any number by, 99, 999, 9999, 99999... involves a pattern in its result. It is necessary to understand the pattern. The readers are advised to note it vigilantly, so that you can evaluate such multiplications within seconds in mind itself.

NOTE: We use the rule of complement or the Vedic sutra, called Nikhilam method to evaluate such multiplications. Recall that, complement / nikhilam is 'all from 9 and last from 10'.

CASE 1: Multiplication by 99 (Nearest base **100**)

#Multiplication of a single digit number by 99

Step 1: Make it into two parts separated by '/' sign. On the right side of '/' write the complement (Nikhilam) of the number from 100.

Step 2: On the left side of '/' write one less than the multiplicand

Example: Evaluate 8 x 99 = 7/92 = 792

Explanation: The complement of 8 from 100 is 92 is written on the right side of '/' and on less than 8, that is 8 − 1 = 7 is written on the left side.

Example: 6 x 99 = 594, 4 x 99 = 396 and so on.

#Multiplication of a two digit number by 99

Step 1: Make it into two parts separated by '/' sign. On the right side of '/' write the complement (Nikhilam) of the number from 100.

Step 2: On the left side of '/' write one less than the multiplicand

Example: 12 x 99= 11/88= 1188 (Complement of 12 is 88 and 12 − 1 = 11)

Example: 27 x 99= 26/73 = 2673 (Complement of 27 is 73 and 27 − 1 = 26)

Example: 97 x 99= 96/04 (Complement of 97 is 04 and 97 − 1 = 96)

CASE 2: Multiplication by 999 (Nearest base **1000**)

#Multiplication of a single digit number by 99

Step 1: Make it into two parts separated by '/' sign. On the right side of '/' write the complement (Nikhilam) of the number from 1000.

Step 2: On the left side of '/' write one less than the multiplicand.

Example: Evaluate 8 x 999 = 7/992 = 7992

The complement of 8 from 1000 is 992. And multiplicand, 7 − 1 = 6.

Example: Evaluate 9 x 999 = 8/991 = 8991

The complement of 9 from 1000 is 991. And multiplicand, 9 − 1 = 8.

Example: Evaluate 3 x 999 = 2997, 4 x 999=3996 and so on.

#Multiplication of a two digit number by 999

Step 1: Make it into two parts separated by '/' sign. On the right side of '/' write the complement of the number from 1000.

Step 2: On the left side of '/' write one less than the multiplicand

Example: 12 x 999= 11/988= 11988 (Complement of 12 from 1000 is 988 and 12 − 1 = 11)

Example: 27 x 999 = 26/973 = 26973 (Complement of 27 from 1000 is 973 and 27 − 1 = 26)

Example: 97 x 999 = 96/903= 96973 (Complement of 97 from 1000 is 903 and 97 − 1 = 96)

Multiplication of a 3 digit number by 999

Step 1: Make it into two parts separated by '/' sign. On the right side of '/' write the complement of the number from 1000.

Step 2: On the left side of '/' write one less than the multiplicand.

Example: 322 x 999= 321/678 = 321988
(Complement of 322 from 1000 is 988 &322−1=321)

Example: 780 x 999= 779/220 = 779220
(Complement of 280 from 1000 is 220& 780 −1= 779)

Example: 998 x 999= 997/002 = 997002
(Complement of 998 from 1000 is 002& 998 −1= 997)

SUMMARY:

I. The multiplication of a single digit or two digit number by 99 can be evaluated directly by using

Fifty Tricks of Multiplication

complement (Nikhilam) rule.

II. The multiplication of a single digit, two digit or a three digit number by 999 can be evaluated directly by using complement (Nikhilam) rule.

III. Extension: Similarly the above problems can be extended to multiplication by any number of 9's. Accordingly, the base shall be changed.

Example: 123456 x 999999 = 123455876544.

NOTE: Complementary rule or Nikhilam sutra is also known as "All from 9 and last from 10".

DO YOURSELF

1. Multiply 4, 8, 93, 94, 29, 91, 63, 57, 45, 88, 76, 89 & 73 by 99.

2. Multiply 5, 6, 43, 94, 129, 891, 323, 513, 245, 888, 767, 809 & 738 by 999.

3. Multiply the above numbers by 9999.

#Tr.29 ANY NUMBER BY 99:

We use the base near by the multiplier, that is 100 split and express 99 as (100 − 1) and evaluate the multiplications by 99.

PROCEDURE: Multiply the multiplicand by 100 and then subtract the number from this product.

Example: 8 x 99 = 8 x (100 − 1) = 800 − 8 = 792.

Example: 12 x 99 = 12 x (100 − 1) = 1200 − 12 = 1288.

Example: 82 x 99 = 82 x (100−1) = 8200 − 82 =

8218.

SHORTCUT: Multiplication by 99 = the number x 100 − the number.

Example: 222 x 99 = 22200 − 222 = 21978.

Example: 765 x 99 = 76500 − 765 = 75735

Example: 3241 x 99 = 324100 − 3241 = 320859

DO YOURSELF

1. Multiply 7, 81, 193, 924, 9129, 9170, 6333, 5757, 41215, & 70003 by 99.

#Tr.30 **ANY NUMBER BY 999:**

We use the base near by the multiplier, that is 100 split and express 999 as (1000 − 1) and evaluate the multiplications by 999.

PROCEDURE: Multiply the multiplicand by 1000 and then subtract the number from this product.

#New Tr. **BINAYAK'S COUPLES:**

(Find pairs of two digit numbers such that **AB x CD = BA x DC**)

We will find series of pairs of two digit numbers when multiplied reversing the numbers gives the same result as the multiplication of original numbers. Let us go for generating possible pairs in series, of such two digit numbers.

Motivation:

12 x 21, 13 x 31, 14 x 41, 15 x 51, - - -

23 x 32, 24 x 42, 25 x 52, - - -

34 x 43, 35 x 53, - - -

SERIES 1: Consider,

12 x 21

24 x 42 (12 & 21 both are multiplied by 2)
36 x 63 (12 & 21 both are multiplied by 3)
48 x 84 (12 & 21 both are multiplied by 4)

We can generate **6** such distinct pairs with different digits of two digit numbers & they are:

12 x 21 = 21 x 12 = 252
12 x 42 = 21 x 24 = 504 – (1)
12 x 63 = 21 x 36 = 756 – (2)
12 x 84 = 21 x 48 = 1008 – (3)
24 x 42 = 42 x 24 = 1008
24 x 63 = 42 x 36 = 1512 – (4)
24 x 84 = 42 x 48 = 2016 – (5)
36 x 63 = 63 x 36 = 2268
36 x 84 = 63 x 48 = 3024 – (6)
48 x 84 = 84 x 48 = 4032

SERIES2: Consider,

13 x 31

26 x 62 (13 & 31 both are multiplied by 2)
39 x 93 (13 & 31 both are multiplied by 3)

We can generate **3** such distinct pairs with different digits of two digit numbers & they are:

13 x 31 = 31 x 13 = 403
13 x 62 = 31 x 26 = 806 – (1)
13 x 93 = 31 x 39 = 1209 – (2)

26 x 62 = 62 x 26 = 1612
26 x 93 = 62 x 39 = 2418 – (3)
39 x 93 = 93 x 39 = 3627

SERIES3: Consider,

23 x 32

46 x 64 (13 & 31 both are multiplied by 2)

We can generate **1** such distinct pair with different digits of two digit numbers & they are:

23 x 32 = 32 x 23 = 736
23 x 64 = 32 x 46 = 1472 – (1)
46 x 64 = 64 x 46 = 2944

SERIES 1: Consider,

34 x 43

68 x 86 (34 & 43 both are multiplied by 2)

We can generate **1** such distinct pair with different digits of two digit numbers & they are:

34 x 42 = 43 x 34 = 1462
34 x 86 = 43 x 68 = 2924 – (1)
68 x 86 = 86 x 68 = 5848

SUMMARY: We have 11 such pairs of two digit numbers with distinct digits, whose multiplication is equal to the multiplication of the same numbers taken in reverse order. (AB x CD = BA x DC).

NOTE: You can generate such pairs of numbers of three digits, that satisfy the Binayak's Couple property, that is **ABC X DEF = CBA X FED**.

HINTS: 123 x 321, 246 x 642, 369 x 963, - - - .

#Tr.31PATTERNS WITH 11:

Pattern 1:

11 x 11 = 121

11 x 111 = 1221
11 x 1111 = 12221

Pattern 2:

11 x 11 = 121
11 x 21 = 231
11 x 31 = 341
11 x 41 = 451 and so on
11 x 81 = 891
11 x 91 = 1001

Pattern 3:

11 x 12 = 132
11 x 22 = 242
11 x 32 = 352
11 x 42 = 462 and so on

Pattern 4:

11 x 99 = 1089
11 x 999 = 10989
11 x 9999 = 109989
11 x 99999 = 1099989 and so on

Pattern 5:

11 x 11 = 121
111 x 111 = 12321
1111 x 1111 = 1234321
11111 x 11111 = 123454321
111111 x 111111 = 12345654321
1111111 x 1111111 = 1234567654321
11111111 x 11111111 = 123456787654321
111111111 x 111111111 = 12345678987654321

CHAPTER FIVE

In this chapter we will learn some interesting tricks those shall deal with the short cut methods of multiplication of any number by any number, irrespective of the number of digits it has.

#Tr.32 VERTICALLY & CROSSWISE:

Vertically & crosswise is otherwise also know as Urdhva – tiryagbhyam in Vedic Mathematics, is the general formula applicable to all cases of multiplication.

Case 1: Multiplication of Two digit numbers. Let the numbers be **ab** and **cd**.

Step 1: Multiply the unit place
Digits, that is b x d = **bd**

Step 2: Make the cross multiplication of both the columns and add them, that is **(ad + bc)**.

Step 3: Multiply the ten's place digits, that is **ac**.
Write them in order as bd/(ad+bc)/ac is the desired product.

NOTE: In any case if the result in any of the above steps exceeds 9, keep the units place and carry the rest of the digits to the left.

Example: Evaluate 23 x 21.
Step 1: 3 x 1 = **3**
Step 2: (2 x 1) + (3 x 2) = **9**
Step 3: 2 x 2 = **4**
The answer is 23 x 21 = **493**

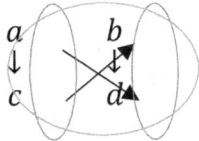

Fifty Tricks of Multiplication

Example: Evaluate 46 x 37.
Step 1: 6 x 7 = **42**
2 is kept and 4 is carried to
The left.
Step 2: (4 x 7) + (6 x 3) = 46
Now 46 + 4 = **50.** O is kept
& 5 is carried to the left.
Step 3: 4 x 3 = 12. Now 12 + 5 = **17**
The answer is 46 x 37 =**1702**

Example: Evaluate 72 x 16.
Step 1: 2 x 6 = **12**
2 is kept and 1 is carried to
the left.
Step 2: (7 x 6) + (2 x 1) = 44
Now 44 + 1 = **45. 5** is kept
&4 is carried to the left.
Step 3: 7 x 1 = 7. Now 7 + 1 = **11**
The answer is 46 x 37 =**1152**

Case 2: Multiplication of Two digit numbers. Let the numbers be **ab** and **cd**.

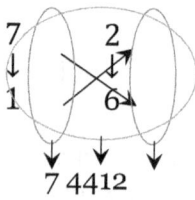

Step 1: Multiply the unit place digits, that is c x f = **cf**
Step 2: Make the cross
multiplication of 1st two columns, and add them, that is **(bf + ce)**.
Step 3: Multiply, following the red directed arrows and add them, that is **(af +be + cf)**.
Step 4: Cross multiply the left two columns & add them, we get, **(ae + bd)**.
Step 5: Multiply the last column digits, & we get, **ad**.
Combine them & write in order as follows:

ad/(ae+bd)/(af+be+cf)/(bf+ce)/cf.

Example: Evaluate 431 x 141.
Step 1: 1 x 1 = **1**
Step 2: (3 x 1) + (1 x 4) = **7**
Step 3: (4 x 1)+(1x1)+(3x4)
= 1**7**. 7 is kept & 1 is carried.
Step 4: (4 x 4) + (3x 1)= 1**9**.
Now, 19 + 1 = 2**0**. 0 is kept and 2 is to be carried.
Step 5: 4 x 1 = 4. Now 4 + 2 = **6**.
So the answer is **60771**.

SUMMARY:
1. For two digit numbers – 3 step multiplication & or addition is done in columns (1st column, both two columns & 2nd column respectively).
2. For 3 digit numbers – 5 step multiplications & or addition is done in columns (1st columns, 1st two columns, all 3 columns, last two columns & last column.

DO YOURSELF

1. Evaluate 77 x 81, 193 x 924, 112 x 342, 917 x 633.

#Tr.33 VERTICALLY & CROSSWISE (4 digit, 5 digit numbers):

This method of vertically & cross wise multiplication can be extended for numbers of any number of digits.
The numbers of steps involved are:
For 2 digit numbers: 3 steps
For 3 digit numbers: 4 steps
For 4 digit numbers: 7 steps
For 5 digit numbers: 9 steps

MULTIPLICATION OF 4 DIGIT NUMBERS

Consider the numbers, abcd & efgh, where the alphabets are any of the digits from 0 to 9.

Step 1: dh
Step 2: (ch + dg)
Step 3: bh + df + cg
Step 4: (ah + de) + (bg + cf)
Step 5: (ag + ce + bf)
Step 6: (af + be)
Step 7: ae

$$X\begin{matrix} a & b & c & d \\ e & f & g & h \end{matrix}$$

Now combine the above numbers to get the final value of the product, abcd x efgh

MULTIPLICATION OF 5 DIGIT NUMBERS

Consider the numbers, abcd & efgh, where the alphabets are any of the digits from 0 to 9.

Step 1: ej
Step 2: (dj + ei)
Step 3: (cj + eh + di)
Step 4: (bj + eg) + (ci + dh)
Step 5: (aj + ef) + (bi + dg) + ch
Step 6: (ai + df) + (bh + cg)
Step 7: (ah + cf + bg)
Step 8: (ag + bf)
Step 9: af

$$X\begin{matrix} a & b & c & d & e \\ f & g & h & i & j \end{matrix}$$

Now combine the above numbers to get the final value of the product, abcde x efghi

NOTE: In case in any step, the sum exceeds 9, keep the unit's place digit & carry the rest of the digits to the left.

DO YOURSELF

1. Evaluate 1727 x 8181, 1053 x 1924, 1312 x 3142, 9617 x 6033, 12345 x 54321, and 67012 x 32141.

#Tr.34 MULTIPLICATION – ANY BASE:

Previously, in Chapter two, we have used the bases of 10, 100, 1000, . . . to evaluate the multiplication of near by numbers of these bases.

Here we will declare an appropriate base for the given numbers to be multiplied & perform the job of multiplications in an easier way. We will try to learn these trick directly through examples. Remember, we will use the multiplication relation method of the declared base with the main base.

Motivation:

I. 50 it 5 x 10

II. 20 is 2 x 10

III. 60 is 6 x 10

IV. 300 is 3 x 100

V. 700 is 7 x 100 and so on.

Example: Evaluate 23 x 22.

We declare a working base 20. And 20 is 2 x 10.

23 – (+3), 3 more than 20

22 – (+2), 2 more than 20
───────────
50 / 6 = 506 {Since, it is (22+3) x 2)/6}

Explanation: 23 is 3 more than 20, So we write +3. 22 is 2 more than 20 & we write +2. Unit's places are multiplied & written on the right part of the '/' sign. On the left part of '/', we write (22 +3) x 2 Or (23 +2) x 2, in both case we get 50. The sum is multiplied by 2, since 20 is 2 x 10.

Example: Evaluate 69 x 76.

We declare a working base 70. And 70 is 7 x 10.

69 – (–1), 1 less than 70

76 – (+6), 6 more than 70

525 /–6 = $525\overline{6}$ = 5244 {Since, it is (69+6) x 7)/–6}

Explanation: 69 is 1less than 70, So we write –1. 76 is 6 more than 70 & we write +7. Unit's places are multiplied & written on the right part of the '/' sign. On the left part of '/', we write (69 +6) x 7 Or (76–1) x 7, in both case we get 525. Now, $525\overline{6}$ is 5250 – 6= 5244. The sum is multiplied by 7, since 70 is 7 x 10.

#Tr.35 MULTIPLICATION – ANY BASE:

We will use the division relation method of the declared base with the main bases.

Motivation:

I. 50 it 100 / 2

II. 25 is 100 / 4

III. 500 is 1000 / 2 and so on.

Example: Evaluate 62 x 48.

We declare a working base 50. And 50 is 100/2.

62 – (+12), 12 more than 50

48 – (–2), 6 more than 50

30 /–24 = $30\overline{24}$ = 2976 {Since, it is (62–2)/ 2) /–6}

Explanation: 62 is 12 more than 50, So we write +12. 48 is 2 less than 50 & we write –2. Unit's places are multiplied & written on the right part of the '/' sign. On the left part of '/', we write (62 –2)/2 Or (48 +12) x 2, in both case we get 30. Now, $30\overline{24}$ is

3000 − 24 = 2976. The sum is divided by 2, since 50 is 100/2.

Example: Evaluate 511 × 497.

We declare a working base 500. And 500 is 1000/2.

511 − (+011), 11 more than 500

497 − (−003), 03 less than 500
----

254 /−033 = $254\overline{0}33$ = 253967

{Since, it is ((511−3)/ 2) /−003} = $254/\overline{0}33$.

Explanation: 511 is 011 more than 500, so we write +011. Here three places are written since, the main base taken is 1000, has three zeros. 497 is 3 less than 500 & we write −003. 012 is multiplied with −003 & we get −033, written on the right part of the '/' sign. On the left part of '/', we write (511−3)/2 Or (497+11) × 2, and in either case we get 254. Now, is $254\overline{0}33$ = 254000 − 33 = 253967. The sum is divided by 2, since 500 is 1000/2.

DO YOURSELF

1. Evaluate 927 × 997, 897 × 1002, 112 × 96, 912 × 1004, and 1130 × 1008.

#Tr.36 MULTIPLICATION OF TWO − LINEAR POLYNOMIALS:

Consider polynomials $ax + b$ and $cx + d$, where a, b, c and d are the coefficients. We can easily evaluate $(ax+b) \times (cx+d)$ using a lattice. Put one polynomial on the top of the table and one on the right hand side of the table. Calculate a product for each cell by multiplying the term

	ax	b	
	acx^2	bcx	cx
	adx	bd	d

at the top of the column and
the term at the right of the row.
Now adding all the columns, we get the value of the
product: (ax +b) x (cx + d) = acx^2 + (ad+bc) x+ bd.
Example: Evaluate (x+2) x (x+3)
By lattice method, we find that the value of the
product is
x^2 + (2+3)x + 6.
= x^2 + (2+3)x + 6.

	x	**2**	
	x^2	2x	**x**
	3x	6	**3**

#Tr.37 MULTIPLICATION OF TWO QUADRATIC EQUATIONS:

Draw a 3 x 3 size table,
since each quadratic
equation has 3 terms.
Write the terms of one
Equation on the top
Of the table aligned
With columns & the
Other one on the right

	ax^2	**bx**	**c**	
	acx^4	$bc\,x^3$	$cd\,x^2$	**dx^2**
	$ae\,x^3$	$be\,x^2$	ce x	**ex**
	$af\,x^2$	bf x	cf	**f**

of the table aligned with the rows. Calculate a
product for each cell by multiplying the term at the
top of the column & the term at the right of the row.
Now adding terms diagonally, we get the value of the
product: $acx^4+(ae+bc)x^3+(af+be+cd)x^2 +(bf+ce)x+cf$

Example: Evaluate $(2x^2 − 4x + 2)$ x $(3x^2 + 5x + 3)$

Draw the 3 x 3 table.
Write $2x^2$, −4x,
and 2 on the top of
the table, aligned
with columns and
write $3x^2$, 5x, and 3,
on the right of the

	$2x^2$	**−4x**	**2**	
	$6x^4$	$−12x^3$	$6x^2$	**$3x^2$**
	$10x^3$	$−20x^2$	10x	**5x**
	$6x^2$	−12x	6	**3**

table, aligned with the rows. Adding the terms diagonally, we get:
$6x^4 - 12x^3 - 8x^2 - 2x + 6$.

#Tr.38 MULTIPLICATION 3 NUMBERS NEAR THE BASES:

Consider the base 10. We will learn it directly through examples.

Example: Evaluate 12 x 13 x 13.
Now, 12 x 13 x 14
= 12^{+2} x 13^{+3} x 14^{+4}. (+2, +3 & +4 represent digits
= 19/26/24   more than 10. 12 is 2 more than
= 2184.   10 etc.)

Explanation:
I. In the 1st part, write (2 x 3 x 4) = 24
II. In the 2nd part, write (2x3 + 3x4 + 2x4) = 26
III. In the 3rd part write (12+3+4) or (13+4+2) or (14+2+3) = 19.
IV. In each part unit place is to be kept and the remaining digit/s to be carried to the left.
NOTE: If the number is more than 10, we write it "+" and if it is less, write "−".

Example: Evaluate 13 x 14 x 8.
13 x 14 x 8
= 13^{+3} x 14^{+4} x 8^{-2}. (−2 since 8 is 2 less than 10)
= 15/−2/(−24)
= 15/−4/−4
= 1456 (1500−44)

Explanation:
I. In the 1st part, write (3 x 4 x −2)) = −24
II. In the 2nd part, write (3x4)+(−2x4)+(−2x3)= −2.
III. In the 3rd part write (13+4+(−2)) or (14+(−2)+3) or (8 + 4 + 3) = 15.

NOTE: Since the base is 10, so we keep only one

place in each part.

Consider the base 100.

Example: Evaluate 104 x 107 x 103.

104 x 107 x 103
= 104^{+4} x 107^{+7} x 103^{+3}. (+4, +7 & +7 represent digits
= 114/61/84   more than 100. 104 is 4 more
= 1146184.   than 10 etc.)

Explanation:
I. In the 1st part, write (4 x 7 x 3) = 84
II. In the 2nd part, write (4x7 + 7x1 + 3x4) = 61
III. In the 3rd part write (104+7+3) or (107+4+3) or (103+4+7) = 114.

NOTE: Since the base is 100, so we keep two digits in each part.

Example: Evaluate 104 x 97 x 103.

104 x 97 x 103
= 104^{+4} x 97^{-3} x 103^{+3}. (−3 since 97 is 3 less from 100)
= 104/−09/−36
= 1039064 (1040000 − 936)

Explanation:
I. In the 1st part, write (4 x 7 x 3) = 84
II. In the 2nd part, write (4x7 + 7x1 + 3x4) = 61
III. In the 3rd part write (104−3+3) or (97+4+3) or (103−3+4) = 104.

NOTE: Since the base is 100, so we keep two digits in each part.

Consider the base 1000.

Example: Evaluate 1005 x 1009 x 1002.

1005 x 1009 x 1002
= 1005^{+5} x 1009^{+9} x 1002^{+2}.
= 1016/073/090 = 1016073090.

Explanation:

I. In the 1st part, write (005 x 009 x 002) = 090
II. In the 2nd part, write (5x9 + 9x2 + 2x5) = 073
III. In the 3rd part write (1005+9+2) or (1009+2+5) or (1002+9+5) = 1016.

NOTE: Since the base is 1000, so we keep three digits in each part.

#Tr.39 MULTIPLICATION OF FOUR NUMBERS NEAR THE BASES:

Consider the base 10. We will learn it directly through examples.

Example: Evaluate 12 x 13 x 14 X 17.
12 x 13 x 14 X 17
= 12^{+2} x 13^{+3} x 14^{+4} x 17^{+7}
= 26/89/206/168
= 37128.

Consider the base 100.

Example: Evaluate 102 x 98 x 97 x 104.
102 x 98 x 97 x 104
= 102^{+2} x 98^{-2} x 97^{-3} x 104^{+4}.
= 101/−16/−04/48
= 100839648.

NOTE: Since the base is 100, so we keep two digits in each part.

#Tr.9 PATTERNS:

Series I:
1 x 9 = 9
12 x 9 = 108
123 x 9 = 1107
1234 x 9 = 11106
12345 x 9 = 111105

123456 x 9 = 1111104
1234567 x 9 = 11111103
12345678 x 9 = 111111102
123456789 x 9 = 1111111101

NOTE: In each of the above products, the sums of the digits equal 9.

Series II:

1 x 9 + 2 = 11
12 x 9 + 3 = 111
123 x 9 + 4 = 1111
1234 x 9 + 5 = 11111
12345 x 9 + 6 = 111111
123456 x 9 + 7 = 1111111
1234567 x 9 + 8 = 11111111
12345678 x 9 + 9 = 111111111
123456789 x 9 + 10 = 1111111111

Series III: (Without 8)

12345679 x 1 x 9 = 111111111
12345679 x 2 x 9 = 222222222
12345679 x 3 x 9 = 333333333
12345679 x 4 x 9 = 444444444
12345679 x 5 x 9 = 555555555
12345679 x 6 x 9 = 666666666
12345679 x 7 x 9 = 777777777
12345679 x 8 x 9 = 888888888
12345679 x 9 x 9 = 999999999

Series IV:

987654321 x 1 x 9 = 888888888 **9**
987654321 x 2 x 9 = **1** 777777777 **8**
987654321 x 3 x 9 = **2** 666666666 **7**
987654321 x 4 x 9 = **3** 555555555 **6**
987654321 x 5 x 9 = **4** 444444444 **5**
987654321 x 6 x 9 = **5** 333333333 **4**

Fifty Tricks of Multiplication

987654321 x 7 x 9 = **6** 222222222 **3**
987654321 x 8 x 9 = **7** 111111111 **2**
987654321 x 9 x 9 = **8** 000000000 **1**

NOTE: In each product the sum of the digits is a multiple of 9.

Series V:

9 x 9 = 81
99 x 99 = 9801
999 x 999 = 998001
9999 x 9999 = 99980001
99999 x 99999 = 9999800001
999999 x 999999 = 999998000001
9999999 x 9999999 = 99999980000001
99999999 x 99999999 = 9999999800000001

NOTE: Sum of the digits in each product is an integer multiple of 9.

#Tr.40 3 DIGIT NUMBERBY 2 DIGIT NUMBER MULTIPLICATION:

We need to multiply a 3 digit number abc by a two digit number de.
The vertically & crow-wise multiplication result shall be as follows:
ad/(ae+bd)/(be+cd)/ce.

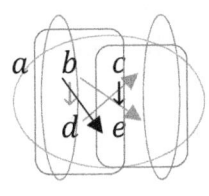

Example: Evaluate 253 x 67.
Following the above rule, we get:
12/(14+30)/(35+18)/21
= 12/44/53/21
=16951. In each part, unit place is kept and the remaining digit is carried to the left.

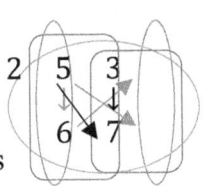

CHAPTER SIX

In this chapter we will learn some additional tricks of multiplication, including to multiply 3 numbers, near to the bases 10, 100, 1000 ... etc. Besides we will note down few patterns, which not only eases our multiplication process but also finds the answer for us within our mind.

#Tr.41 MULTIPLICATION – ANY BASE:
Working with base 10:
Example: 7 x 7. The nearest base is 10.
Using the method: "All from 9 and the last from 10":
7 − 3 (3 less than the base 10)
x 7− 3 (3 less than the base 10)

49

Left side: 7 + 7 - 10= 4 and right side: (−3) x (−3) = 9
The answer is (4 x 10) + 9 = 40 +9 = 49
Example: 18 x 14. The nearest base is 10.
Using the method: "All from 9 and the last from 10":
  18 +8 (8 more than the base 10)
x 14+4 (4 more than the base 10)
---
  22 32

Left side: 18+14−10= 22and right side: 8 x 4 = 32.
The answer is (22 x10) + 32 = 2200 + 32 = 2232
Working with base 100:
Example: 108 x 98. The nearest base is 100.
Using the method: "All from 9 and the last from 10":
   108 +8 (8 more than the base 100)
x 98 −2 (2more than the base 100)

106−16

Left side: 108+ 98−100= 106and
right side: 8 x (−2) = −16.

The answer is (106 x 100) −16 = 10600 −16 = 10584.

Working with base 1000:

Example: 1034 x 1004.
Using the method: "All from 9 and the last from 10":
 1034   +34 (34 more than the base 1000)
x 1002   +2 (2 more than the base 100)
 ───────────
  366   68
Left side: 1034 + 1002 − 1000 = 1036 and
right side: 8 x (−2) = −16.
The answer is (1036 x 1000) + 68
= 1036000 − 68 = 1036068.

#Tr.42 MULTIPLE OF 7, 11 & 13:

It is interesting to note that, the 3 consecutive prime numbers, namely, 7, 11 & 13 multiplied together gives 1001. That is 7 x 11 x 13 = 1001.

Trick: Any 3 digit number multiplied by 1001 = the three number, written two times is the answer.

Example:
1001 x 123 = 123123
1001 x 789 = 789789
1001 x 234 = 234234 etc.

Let us explore more use of it through examples:

Example: Evaluate 234 x 91 x 11.
We can re-write 234 x 91 x 11 as 234 x 7 x 13 x 11
= 234 x 1001 = 234234.

Example: Evaluate 903 x 77 x 13.
We can re-write 903 x 77 x 13 as 903 x 7 x 11 x 13
= 903 x 1001 = 903903.

NOTE: If we find the combination of 7, 11, & 13 multiplied together, the above trick can be applied.

Example: 28 x 143 x 223.
Using IQ, we can explore that 7 hidden inside 28.
Now, 28 x 143 x 223 can be written as:
4 x 7 x 11 x 13 x 223
= 4 x 1001 x 223
= 4 x 223223
= 2 x 446446
=492492.

Trick: Any 4 digit number, say **ABC** multiplied by 1001 = **ABC(A+D)BCD**.

Note: In case, the sum total (A+D) exceeds 9, retain the first digit and carry over the other digits to the left.

Example: Evaluate 2345 x 7 x 11 x 13.
2345 x 7 x 11 x 13 = 2345 x 1001 = 1237345.
NOTE: Here A + D is 7.

DO YOURSELF

1. Evaluate 14 x 101 x 11 x 13, 105 x 55 x 91, 13 x 77 x 987, 99 x 7 x 143, and 77 x 13 x 214.

#Tr.43 **MULTIPLICATION BY USING FACTORING:**

Motivation: Often everyone enjoys single digit as a multiplier. Keeping this in interest, we can try easier way to evaluate the multiplications in mind.

Here, factoring means breaking a given number into single digit factors.

Example: 24 can be factored as 2 x 4 x 3 or 8 x 3 or 2 x 2 x 2 x 3 or 6 x 4 etc.

Example: 72 x 42.
It can be 72 x 7 x 6 = 504 x 6 = 3024
Explanation: 72 x 7 = 70 x 7 + 2 x 7 = 504.

Fifty Tricks of Multiplication

Example: 85 x 63.
It can be 85 x 9 x 7 = 755 x 7 = 5285
Explanation: 85 x 9 = 80 x 9 + 5 x 7 = 755.
755 x 7 = 750 x 7 + 5 x 7 = 5285
Example: 46x42=46x(7x6)=(46x7)x6=322x6=1932
Example: 57x24=57x(4x6)=(57x4)x6=228x6= 1368
Example: 57x24=57x(6x4)=(57x6)x4=342x4= 1368
Example: 57x24=57x(8x3)=(57x8)x3=456x3= 1368

NOTE: So, it is not unique, but, there are some familiar products, that may really help us to some extent. Suppose we go sequentially, it is as follows:
12 x 9 = 108
13 x 8 = 104 = 26 x 4 = 52 x 2, 26 x 8 = 208 = 52 x 4
15 x 9 = 105, 21 x 5 = 105, 35 x 3 = 105,
17 x 6 = 102 = 34 x 3 = 51 x 2, 34 x 6 = 204 = 51 x 4,
68 x 3 = 204, 68 x 6 = 408
18 x 6 = 108, 27 x 4 = 108, 36 x 3 = 108, 54 x 2 = 108
23 x 9 = 207
25 x 4 = 100, 25 x 8 = 200
29 x 7 = 203
34 x 9 = 206
38 x 8 = 304
41 x 4 = 205
43 x 7 = 301
44 x 7 = 308, 77 x 4 = 308
45 x 9 = 405
51 x 6 = 306, 51 x 8 = 408
53 x 2 = 106
56 x 9 = 504 = 84 x 6
61 x 5 = 305
63 x 8 = 504 = 84 x 6
67 x 3 = 201, 67 x 6 = 402, 67 x 9 = 603
72 x 7 = 504
76 x 4 = 304, 76 x 8 = 608

78 x 9 = 702
81 x 5 = 405
88 x 8 = 704
89 x 9 = 801

#Tr.44 MULTIPLICATION OF DECIMAL NUMBERS:

Case I: Normal decimal multiplications

Sep 1: Do the multiplication ignoring the decimals (treating the numbers as integers). Count the number of decimal places in each number, let they are m &n.

Step 2: In the product put the decimal (dot) sign after (m + n) places from the right towards left.

Example: Evaluate 1.12 x 0.108. 1.12 has two and 0.108 has 3 decimal places.
We multiply 112 x 108 = 120/96= 12096.
Now put the decimal (dot) sign after (2+3=5) places from the right. The answer is 1.12 x 0.108 = 0.12096

Example: Evaluate 1.12 x 0.25.
1.12 has two and 0.25 has 3 decimal places.
We multiply 112 x 25 = 2800. (Recall from earlier chapter that, multiplication by 25 is same as halving 112 2 times and multiply by 100)
Now put the decimal (dot) sign after (2+2=4) places from the right. The answer is 1.12 x 0.25 = 0.2800.

DO YOURSELF

1. Evaluate 0.14 x 0.50, 21.05 x 305, 11.3 x 11.2, 9.97 x 10.4, and 7.7 x 1.3.

Case II: When the integer part is same and the decimal parts add to 1.
Step 1: On the right side multiply the digits as we did in case I.

Step 2: In the left side multiply the digit/s with one more than the previous one.

Example: Evaluate 3.25 x 3.75.

Here the integer part is common, which is 3 & the decimal parts add to 1.

3.25 x 3.75 = (3 x 4)/ (0.25 x 0.75)

= 12 / 0.1875 = 12.1875

Example: Evaluate 9.5 x 9.5.

9.5 x 9.5 = (9 x 10)/ 0.5 x 0.5) = 90/0.25= 90.25

Example: Evaluate 7.46 x 7.64.

7.46 x 7.64. = (7 x 8)/ (0.46 x 0.64) = 56.2944

Case III: Mixed fraction multiplication

Example: 7 ¼ x 7 ¼ = (7 x 8) / (¼ x ¼) = 56 ¼.

NOTE: we used one more than the previous one.

#Tr.45 MULTIPLICATION NUMBERS, LAST TWO DIGITS ADDS TO 100:

Consider the base 100.

Example: Evaluate 192 x 108.

```
  192   +92  (92 more than 100)
x 108   +08  (8 more than 100)
―――――――――――――――――――――――――
  2 / 0736 = 20736  (1 x 2 = 2 and 92 x 08 = 0736)
```

Explanation: Here 92 & 08 adds to 100. The hundred's place is same. We use, one more than the previous one for the left part of '/' and it is 1 x 2. IN the right side it is 92 x 08 = 0736.

Fifty Tricks of Multiplication

Example: Evaluate 413 x 487.

```
  413   +13  (13 more than 400)
x 487   +87  (87 more than 400)
```
20 / 1131 = 201131 (4 x 5 = 20 and 13 x 087 = 1131)

Explanation: Here 92 & 08 adds to 100. The hundred's place is same. We use, one more than the previous one for the left part of '/' and it is 1 x 2. IN the right side it is 92 x 08 = 0736.

DO YOURSELF

1. Evaluate 345 x 355, 215 x 285, 113 x 187, 997 x 903, and 727 x 777.

#Tr.46 USE OF ONE LESS THAN THE PREVIOUS / ONE LESS THAN THE ONE BEFORE: ALTERNATIVE USE

Case I: Multiplication by 999

We use the steps as follows:

Step 1: On the left, write one less than the given number

Step 2: Subtract the result from the multiplier.

Example: Evaluate 672 x 999.

Using 1 less than the previous one on the left side and subtract the result from 999 for the right side..

672 x 999 = (672-1)/(999-(672-1)) = 671328

Example: Evaluate 84 x 9999.

Using 1 less than the previous one on the left side and subtract the result from 9999 for the right side..

84 x 9999 = (84-1)/(9999-(84-1)) = 839916

Example: Evaluate 84 x 99999.

Using 1 less than the previous one on the left side and subtract the result from 99999 for the right side..

84 x 99999 = (84-1)/(99999− (84-1)) = 8399916

Example: Evaluate 675 x 99.

Using 1 less than the previous one on the left side, write 99 on the left and then subtract the left side from the total number.

675 x 99

= (675−1)/(99) − (675−1) = 67499 − 674 = 66825.

Example: Evaluate 6751 x 999.

Using 1 less than the previous one on the left side, write 999 on the left and then subtract the left side from the total number.

6751 x 999
= ((6751−1)/(999)) − (6751−1)
= 6750999 − 6750 = 6744249.

Example: Evaluate 72 x 9.

72 x 9 = 71/9 − 71= 719 − 71 = 648.

DO YOURSELF

1. Evaluate 44 x 9, 44 x 99, 2105 x 99, 113 x 999, 997 x 9, and 717 x 99999.

#Tr. **MULTIPLICATION BY 9'S, A SYSTEMETIC ANALYSIS:**

I. When the multiplicand and multiplier (9's) both have the same number of digits

Fifty Tricks of Multiplication

Example: 8 x 9, 21 x 99, 234 x 999, 1234 x 9999, etc.

II. When the multiplier has more number of digits (9's) than the multiplicand.

Example: 7 x 99, 12 x 999, 437 x 9999 etc.

II. When multiplierhas lesser digits (9's) than the multiplicand

Example: 23 x 9, 543 x 99, 6771 x 999, etc.

We will conclude for a rule, after observing the following series of multiplications.

OBSERVATION: The 9 table

Series I

11 x 9 = 9/9 = 99
12 x 9 = 10/8 = 108
13 x 9 = 11/7= 117---
18 x 9 = 16/2 = 162
19 x 9 = 17/1 = 171
20 x 9 = 18/0*= 180

Series II

21 x 9 = 18/9 = 189
22 x 9 = 19/8 = 198
23 x 9 = 20/7= 207---
28 x 9 = 25/2 = 252
29 x 9 = 26/1 = 261
30 x 9 = 27/0 = 270

Series III

31 x 9 = 27/9 = 279- - -
35 x 9 = 31/5 = 315
36 x 9 = 32/4= 324 - - -
46 x 9 = 41/4 = 414
53 x 9 = 47/7 = 477
67 x 9 = 60/3= 603---so on.

From the series detailed above, the following points can be observed:

Observation 1:
Series I has the multiplicands with 1 as first digit except the last one, which is 2. Here, left hand side of the products are uniformly 2 less than the multiplicands.
*The same is also true with20 x 9.

Observation 2:
Series II has the same pattern. Here L.H.S of products are uniformly 3 lessthan the multiplicands.

Observation 3:

Series III is of mixed examples and yet the same result i.e. if 3 is first digit of the multiplicand then L.H.S of product is 4 less than the multiplicand; if 4 is first digit of the multiplicand then, L.H.S of the product is 5 less than the multiplicand and so on.

Observation 4:
The right hand side of the product in all the series of multiplications, cases is obtained by subtracting the R.H.S. part of the multiplicand by Nikhilam.

Conclusion:
We conclude the following steps to evaluate the multiplications, where the multiplier has number of 9's less than the number of digits of the multiplicand.

Step 1: Put a separator ('/')in the multiplicand after the digits from the right equal to the number of 9's in the multiplier.

Step 2: Subtract from the multiplicand one more than the whole excess part on the left.

Step 3: The answer is (Multiplicand −result in step2)/ complement (Nikhilam) of the left part of separator.

Let us now try the following examples:

Example: Evaluate 31 x 9
Step 1: 3/1 (separator is put after 1 digit from the right)
Step 2: Left part of 31 is 3. Now 3 + 1 = 4
We calculate 31 − 4 = 27.
Step 3: The answer is (31−4)/(complement of 1)
= 279

Example: 135 x 9. It is written as 13/5.
The answer is (135 − 14)/5 = 1215

Example: 25637 x 99
Since the multiplier has 2 digits, the answer is
[25637 − (256 + 1)] / (100 − 37)
= (25637 − 257) / 63 = 2538063.

#Tr.47 THE FIRST BY THE FIRST AND THE LAST BY THE LAST:

The problems related to area or volumes of a certain region are associated with the measurements in feet and or inches. For example the area of a rectangular region is the multiplication of length and the breadth of that region.
Area = length x breadth.

Example: Find the area, when length = 7' 4" (7 ft. & 4 inches) and breadth = 3' 8" (3 ft. & 5 inches).

Method I:
Area = Length X Breath
= 7' 4" x 3' 8", and 1' = 12",
= (7 X 12 + 4) (3x 12 + 8) in to single unit
= 88" x 44" = 3872 Sq. inches.
Since 1 sq. ft. =12 X 12 = 144sq.inches we have area equals 3872 / 144 = 26 sq. feet 128 sq. inches.

Method II:

By Vedic principles we proceed in the way "the first by first and the last by last".
i.e. 7' 4" can be treated as $7x + 4$ and 3' 8" as $3x + 8$,
Where x= 1ft. = 12 in; and x^2 is sq. ft.
Now Area = Length X Breath
= 7' 4" x 3' 8", and 1' = 12",
= $(7x + 4)(3x + 8)$
= $21x^2 + 7 \times 8x + 4 \times 3x + 32$
= $21x^2 + 56x + 12x + 32$
= $21x^2 + 68.x + 32$
= $21x^2 + (5x + 8).x + 32$, writing $68 = 5 \times 12 + 8$
= $26x^2 + 8x + 32$
= 26 Sq. ft. + 8 x 12 Sq. in + 32 Sq. in
= 26 Sq. ft. + 96 Sq. in + 32 Sq. in
= 26 Sq. ft. + 128 Sq. in

Method III:

We will use the vertically and cross-wise multiplication.

$$\begin{array}{r} 7'\ 4" \\ \times\ 3'\ 8" \\ \hline \end{array}$$

21/68/32 (7 x 3 = 21/7 x 8 + 4 x 3 =68/4 x8=32)
= 21/ 5 x 12 + 8/ 32 (60 = 5 x 12 +8)
= 26 / 8 / 32   (5 is carried to left)
= 26/8 x 12/32
= 26/128
= 26 sq. ft. 128 sq. inches

DO YOURSELF:

1. Find the area of the rectangles in each of the following situations.

A. l = 4'3", b = 3'4 "
B. l = 11'3", b = 5'5"

C. l = 4 yard 2 ft. b = 1 yards 5 ft.(1yard =3ft)
D. l = 12 yard 2 ft. b = 4 yards 1 ft.

#TR.48 MULTIPLICATION FACTS:

#I

When we multiply 6 by any even number, 2, 4, 6, 8
a. then both end in the same digit.
b. the number in the ten's place will be half of the number in the unit's place.
Example: 6x**2** =1**2**, 6x**4** =2**4**, 6x**6** =3**6**, 6x**8** =4**8**

#II

There is a multiplication twin all the time & it is obvious. For example, 2 x 8 = 16 is same as 8 x 2. So it is existing in the multiplication table.

#III

Think about the digits, 5, 6, 7 & 8, how closely are they related, 56 = 7 x 8.

#IV

Three digits added together and multiplied together, results same, are 1, 2 & 3. 1 + 2 + 3 = 1 x 2 x 3.

#V

Square of numbers with 5 in the unit place, uses the rule of 'One more than the previous one' & the result is as follows:

15^2 = 15 x 15 = 1 x 2 / 5 x 5 = 2/25 = 225

25^2 = 25 x 25 = (2 x 3)/25 = 625

35^2 = 35 x 35 = (3 x 4)/25 = 1225, - - -

95^2 = 95 x 95 = (9 x 10)/25 = 9025 & so on.

Square of a number comes under multiplication.

My new book: "Square, Square Root; Cube & Cube Root in One Day" is available on e-book stores.

#VI.

Multiply: **(x + 5)(x + 7)**.

We have conducted such multiplications by lattice method, earlier.

Now, we will use the vertically and cross-wise method to evaluate it. This means that the x and the 3 in x+3 must both multiply the x and the 4 in x+4. We write these as follows:

x   +   5
x   +   7
─────────────
x^2 + (7x +5x) + 35 (vertical/cross-wise/vertical)
= x^2 + 12x + 35.

Multiply vertically on the right: 7×5 = **35**.

Multiply cross-wise: 7x + 5x = **12 x**

Multiply vertically on the: $x \times x = x^2$.

#Tr.49 NEW NOTATION: 2 DIGITS

Step 1: Write ab x cd as in the figure
Step 2: Write ac/(ad +bc)/bd.
That is, (1)/(2)+(3)/(4).

Example: Evaluate 72 x 41.
72 x 41 = 28/(7+8)/2 = 28/15/2= 2952 (1 is carried from 15).

Example: Evaluate 64 x 37.
64 x 37 = 18/(42+12)/28
= 18/54/28
= 2368 (2 is carried from 28 & 5 is carried rom 54).

Example: Evaluate 91 x 83.
91 x 83 = 72/(27+8)/3

= 72/35/3
= 7553 (3 is carried from 35).

DO YOURSELF

1. Evaluate 54 x 49, 94 x 37, 74 x 87 and 43 x 49.

#Tr.50 SPLITTING TECHNIQUE OF MULTIPLICATION:

We split the digits in the multiplicand suitably, to ease/ quicker our multiplication.

Example:
454 x 2 = 45 / 4 x 2 = 908
604 x 9 = 6 / 04 x 9 = 5436
432 x 3 = 4 / 32 x 3 = 1296
1234223 x 2 = 12 / 34 / 223 = 2468446
515 x 5 = 5 / 15 x 5 = 2575
3193 x 3 = 3 / 19/ 3 x 3 = 9579
1416 x 4 = 14 / 16 x 4 = 5664

DO YOURSELF

1. Evaluate 154 x 3, 293 x 3, 272 x 4 and 243 x 3.

#Tr.51 DIGIT SUM CHECK OF THE MULTIPLICATIONS:

#I

Digit sum is the sum of the digits of a number & we go on adding the digits unless we reach at a single digit.

Step 1: Calculate the digit sum of the numbers to be multiplied. Calculate the product of these digit sums & find the **digit sum of the result**.

Step 2: Calculate the digit sum of the product of the numbers. If the digit sum in step 1 and step 2 are equal, then the answer is correct.

Example: 23 x 27 = 621

Step 1: The digit sum of 23 is 2 + 3 = 5
The digit sum of 27 is 2 + 7 = 9
The multiplication of digit sums 5 and 9 is 45. Now the digit sum of 45 is 4 + 5 = **9**
Step 2: The digit sum of the product 621 is 6 + 2 +1 = 9. The digit sum in step 1 = Digit sum in Step 2.
Example: Check the digit sum in 91 x 83 = 7553.
Step 1: The digit sum of 91 is 9 + 1 = 10, 1 + 0 =1
The digit sum of 83 is 8 + 3 = 11, 1 + 1 = 2
The multiplication of digit sums 1 and 2 is 2.
Step 2: The digit sum of the product value,
7553 is 7+5+5+3 =**20,** 2+0 = **2**
The digit sum in step 1 = Digit sum in Step 2.

#II

Consider the Multiplication of $(x + 5)(x + 7)$.

The algebraic form of the digit sum can also be thought of to be checked as follows:
We have: $(x + 5)(x + 7) = x^2 + 12x + 35$
We check that the product of the sum of the coefficients in the brackets on the left-hand side equals the sum of the coefficients on the right-hand side.
That is $(1 + 5)(1 + 7) = 1 + 12 + 35 = (48)$

Since both sides come to 48 this confirms the answer.

ABOUT THE AUTHOR

Dr. Binayak Sahu:

The author is a highly experienced professor of Mathematics with more than 27 years of teaching, research & administrative experience. He has the specialization in mapping students mind & their learning attitude, taken all into analysis in preparing this book. He has taught Mathematics at various levels starting from higher secondary to Post Graduates across B.Tech. BCA, and other UG students.

He is well qualified with M.Sc., M.Phil., Ph.D. (Mathematics), DIT, DCA – CEDTI (GoI) &LL.B as his degrees / Certificates. He has already authored two books namely, Engineering Mathematics – I, II. He is the **life member of** Indian Society of Theoretical &Applied Mechanics, IIT Kharagpur; SAE India, Chennai; Indian Science Congress Association, New Delhi; ISRD; INAEG; Indian Mathematical Society, Aurangabad. He has various other publications in journals, editorials and books.

www.ingramcontent.com/pod-product-compliance
Lightning Source LLC
Chambersburg PA
CBHW051540240526
45465CB00028B/1625